Manuel Engelbauer

Global assessment of recent UNESCO Biosphere Reserve quality enhancement strategies and interlinkages with other UNESCO labels

Manuel Engelbauer

Global assessment of recent UNESCO Biosphere Reserve quality enhancement strategies and interlinkages with other UNESCO labels

Würzburg University Press

Dissertation, Julius-Maximilians-Universität Würzburg
Philosophische Fakultät, 2022
Gutachter: Prof. Dr. Hubert Job, Prof. Dr. Heiko Paeth

Impressum

Julius-Maximilians-Universität Würzburg
Würzburg University Press
Universitätsbibliothek Würzburg
Am Hubland
D-97074 Würzburg
www.wup.uni-wuerzburg.de

Print on Demand

Coverdesign: Holger Schilling

ISBN 978-3-95826-196-9 (print)
ISBN 978-3-95826-197-6 (online)
DOI 10.25972/WUP-978-3-95826-197-6
URN urn:nbn:de:bvb:20-opus-286538

Acknowledgements

The accomplishment of this doctoral dissertation is linked to the COVID-19 pandemic. While being forced to work from a home office over the last two years has been instrumental in finalising this dissertation, the pandemic also led to new travel restrictions. Therefore, looking back, I am first and foremost grateful for the opportunity to do empirical research that involved fieldwork interviewing experts on all continents – sometimes giving me a feeling of being trapped in a Jules Verne novel. This was such a meaningful experience that not only shaped my worldview but also defined my career path. Through discussions with experts in Beijing, Moscow, and Washington D.C. – despite the current political world situation – I met strong advocates of nature conservation, which strengthens my firm opinion that the exchange of people and cultures will contribute to peaceful coexistence globally and joint solutions to the most urgent global crises: biodiversity loss, climate change and pollution.

The successful conclusion of this dissertation would not have been possible without the endorsement of a number of people, all of whom I would like to express my appreciation at this point.

First and foremost, my gratitude and appreciation goes to my 'Doktorvater' Univ.-Prof. Dr. Hubert Job, without whom this academic path would not have been possible. Exactly 11 years ago, our professional and personal lives were joined on an excursion to Kenya and Tanzania. In the partly 'uncontrolled' time, we not only went on three more joint excursions through Baja California, Central America and Southern Africa but also travelled thousands of kilometres together through the remotest corners of Germany, Europe and Namibia – always in search of the most rustic place to stay overnight and the most majestic species of flora and fauna. Dear Hubert, I am deeply grateful for your constant support, your worldview, your faith in my abilities, your academic guidance, your passion for nature conservation and your perseverance over the last few months to finish this doctoral thesis, crowning our academic life together.

I also wish to express my sincere gratitude to my secondary assessor for this thesis, Univ.-Prof. Dr. Heiko Paeth, for his support. Additionally, I extend my genuine thanks to Prof. Dr. Ralf Klein and Dr. Ludger Brenner for their academic supervision and support as part of my mentorship.

Another heartfelt thank you goes to my academic companions during my seven years at the Chair of Geography and Regional Research at the University of Würzburg: Prof. Dr. Marius Mayer, Dr. Felix Kraus, Dr. Julius Arnegger, Dr. Manuel Woltering, Dr. Ferdinand Paesler, Dr. Anu Lama, Dr. Daniel Mann, Dr. Cornelius Merlin, Dr. Johannes Schamel, Dr. Joachim Rathmann, Niklas Scheder, Dr. Philipp Sacher, Dr. Lisa Majewski, Sarah Bittlingmaier and Winfried Weber. A very special thank you to Christoph Dubrow and Matthias Barrett, who were not only a great support professionally but also always had my back personally. I would also like to thank Karin Menz for her constant support during our almost ten years together at the University of Würzburg.

This study would not have been possible without the financial support of the Research Fund of the Faculty of Arts (Historical, Philological, Cultural and Geographical Studies) at Julius-Maximilians-University Würzburg, which supported the field work I conducted.

A special thanks also goes to the experts who participated in interviews for this dissertation – they constantly conveyed their passion for engagement in the design and implementation of the UNESCO Man and the Biosphere Programme and inspired me deeply.

Last but not the least, I owe sincere gratitude to my friends, parents, and family, who have always supported me, despite many absences and a seemingly never-ending workload. Only through your support and constant allegiance was it possible for me to accomplish this work. Particularly, I would like to thank my trusted friends, who have been there in good times and bad, Susanne Smolorz, Annalena Oppel, Kristina Stubenrauch and Verena Zull. I would like to thank my beloved partner, Mareike Rappold, from the bottom of my heart for her constant support and love during these challenging last few years. And last but not least, I would like to thank my dear parents, Margit and Werner Engelbauer, for their trust and tireless support of my education. Without you it would not have been possible to write this thesis.

Table of contents

List of Figures

List of Tables

Abbreviations

BfN	German Federal Agency for Nature Conservation
CBD	Convention on Biological Diversity
EU	European Union
GIS	Geographic Information System
GGN	Global Network of National Geoparks
GRULAC	Group of Latin American and Caribbean Countries
HAC	High Ambition Coalition
HDI	Human Development Index
IUCN	The International Union for Conservation of Nature
MAB	Man and the Biosphere Programme
MAB IAC	International Advisory Committee for Biosphere Reserves
MAB ICC	International Coordination Council of the MAB Programme
MAB ISG	International Support Group to the MAB Programme
MAP	Madrid Action Plan
MDGs	Millennium Development Goals
MIDAs	Multi Internationally Designated Areas
MUKE	Ministerium für Umwelt, Klima und Energiewirtschaft / Ministry for the Environment, Climate and Energy
NATCOM Kenya	Kenyan National Commission for UNESCO
NGS	National Geographic Society
RERB	Red Española de Reservas de la Biosfera / Spanish Network of Biosphere Reserves
SDGs	Sustainable Development Goals
UN	United Nations
UNDP	United Nations Development Programme
UNEP	United Nations Environment Programme
UNEP-WCMC	UNEP World Conservation Monitoring Centre
UNESCO WHC	UNESCO World Heritage Centre
UNESCO	United Nations Educational, Scientific and Cultural Organization
WBGU	Wissenschaftlicher Beirat der Bundesregierung Globale Umweltveränderungen
WDPA	World Database on Protected Areas
WHS	World Heritage Site
WWF	World Wide Fund for Nature

Abstract

In 1995, the Second International Biosphere Reserve Congress in Seville resulted in a set of new regulations that spurred a significant paradigm shift in the UNESCO Man and Biosphere (MAB) Programme, reconceptualizing the research programme as a modern instrument for the dual mandate of nature conservation and sustainable development. But almost 20 years later, a large proportion of biosphere reserves designated before 1996 still did not comply with the new regulations. In 2013, the International Coordination Council of the MAB Programme announced the 'Exit Strategy' to assess, monitor and improve the quality of the World Network of Biosphere Reserves. However, the strategy also meant that 266 biosphere reserves in 76 member states were faced with the possibility of exclusion from the world network.

This study presents a global assessment of the challenges that result from the Exit Strategy and the Process of Excellence and Enhancement that follows. Specifically, it investigates the differences in quality management strategies and the periodic review processes of various biosphere reserves, the effects of those quality management strategies on the MAB Programme and on the 76 directly affected member states, and the interlinkages between the MAB Programme and other UNESCO designations for nature conservation: the natural World Heritage Sites and the Global Geoparks.

Semi-structured expert interviews were conducted with 31 participants in 21 different countries, representing all UN regions. To showcase the diversity of the World Network of Biosphere Reserves, 20 country-specific case studies are presented, highlighting the challenges of implementing the biosphere reserve concept and, more specifically, the periodic review process. Information gleaned from the experts was transcribed and evaluated using a qualitative content analysis method.

The results of this study demonstrate major differences worldwide in the implementation biosphere reserves, especially in the case of the national affiliation of the MAB Programme, the legal recognition of biosphere reserves in national legislation, the usage of the term 'biosphere reserve' and the governance structures of the biosphere reserves. Of those represented by the case studies, the four countries with the highest number of voluntary biosphere reserves withdrawals after 2013, Australia, Austria, Bulgaria and the United States of America, show that the Exit Strategy contributed to the streamlining and quality enhancement of the world network. The biosphere reserves in those countries were strictly nature conservation areas without human settlements and were designated as such in the 1970s and 1980s. Only post-Seville biosphere reserves remain in those countries. Some experts have pointed out that there appears to be competition for political attention and funding between the three UNESCO labels for nature conservation. While a combination of the designation of biosphere reserves and World Heritage Sites in one place is favoured by experts, Global Geoparks and Biosphere Reserves are seen as being in competition with each other.

This study concludes that quality enhancement strategies were fundamental to improving the credibility and coherence of the MAB Programme. Most pre-Seville biosphere reserves were adapted or the member states were encouraged to withdraw them voluntarily.

Challenges in implementing the Exit Strategy were not unique to individual countries but applied equally to all member states with pre-Seville sites. Over the course of the quality enhancement process, many UNESCO member states have become more involved with the MAB Programme, which has led to rejuvenation of the national biosphere reserves network in many countries.

Zusammenfassung

Im Jahr 1995 führte der zweite internationale Kongress für Biosphärenreservate in Sevilla zu einer Reihe neuer Richtlinien, die einen bedeutenden Paradigmenwechsel im UNESCO-Programm „Der Mensch und die Biosphäre" (MAB) einleiteten und das bestehende Forschungsprogramm in ein modernes Instrument für das doppelte Mandat des Naturschutzes und der nachhaltigen Entwicklung entwickelte. Doch fast 20 Jahre später entsprach ein großer Teil der vor 1996 ausgewiesenen Biosphärenreservate immer noch nicht den neuen Vorschriften. Im Jahr 2013 verkündete der Internationale Koordinierungsrat des MAB-Programms die „Exit-Strategie" zur Evaluierung, Monitoring und Qualitätsverbesserung des Weltnetzes der Biosphärenreservate. Die Exit-Strategie bedeutete jedoch auch, dass 266 Biosphärenreservate in 76 Mitgliedsstaaten mit der Möglichkeit des Ausschlusses aus dem Weltnetz konfrontiert wurden.

Diese Studie präsentiert eine globale Bewertung der Herausforderungen, die sich aus der Exit-Strategie und dem darauffolgenden Prozess der Exzellenz und Aufwertung ergeben. Es werden insbesondere die Unterschiede in den Qualitätsmanagementstrategien und den periodischen Überprüfungsprozessen der verschiedenen Biosphärenreservate, die Auswirkungen dieser Qualitätsmanagementstrategien auf das MAB-Programm und auf die 76 direkt betroffenen Mitgliedsstaaten sowie die Verflechtungen zwischen dem MAB-Programm und anderen UNESCO-Naturschutzsiegeln untersucht: die Weltnaturerbestätten und die Globalen Geoparks.

Es wurden halbstrukturierte Experteninterviews mit 31 Teilnehmern aus 21 verschiedenen Ländern geführt, die alle UN-Regionen repräsentieren. Um die Vielfalt des Weltnetzes der Biosphärenreservate zu veranschaulichen, werden 20 länderspezifische Fallstudien vorgestellt, in denen die Herausforderungen bei der Umsetzung des Biosphärenreservatskonzepts und insbesondere des periodischen Überprüfungsprozesses beleuchtet werden. Die von den Experten gesammelten Informationen wurden transkribiert und mit Hilfe einer qualitativen Inhaltsanalyse ausgewertet.

Die Ergebnisse dieser Studie zeigen, dass es weltweit große Unterschiede bei der Implementierung von Biosphärenreservaten gibt, insbesondere was die nationale Zuständigkeit für das MAB-Programm, die rechtliche Verankerung von Biosphärenreservaten in der nationalen Gesetzgebung, die Verwendung des Begriffs „Biosphärenreservat" und die Governancestrukturen der Biosphärenreservate betrifft. Von den Fallbeispielländern dieser Arbeit zeigen die vier Nationen mit den meisten freiwilligen Rücknahmen von Biosphärenreservaten aus dem Weltnetzwerk nach 2013, nämlich Australien, Österreich, Bulgarien und die Vereinigten Staaten von Amerika, dass die Exit-Strategie zur Vereinheitlichung und Qualitätsverbesserung des Weltnetzes beigetragen hat. Die Biosphärenreservate in diesen Ländern waren reine Naturschutzgebiete ohne menschliche Besiedlung und wurden in den 1970er und 1980er Jahren als solche ausgewiesen. In diesen Ländern gibt es nur noch Biosphärenreservate, die den Qualitätsstandards nach der Konferenz von Sevilla im Jahr 1995 entsprechen. Einige Experten haben darauf hingewiesen, dass es zwischen den drei UNESCO-Naturschutzsiegeln einen Wettbewerb um politische Aufmerksamkeit und Finanzierung gibt. Während eine Kombination von Biosphärenreservaten und Weltnaturerbestätten an einem Ort von Experten favorisiert wird, werden Globale Geoparks und Biosphärenreservate als miteinander konkurrierend angesehen.

Diese Arbeit kommt zu dem Schluss, dass die eingeführten Strategien zur Qualitätsverbesserung von grundlegender Bedeutung waren, um die Glaubwürdigkeit und Kohärenz des MAB-Programms zu verbessern. Die meisten Biosphärenreservate aus der ersten Generation vor der Sevilla-Konferenz wurden angepasst oder die Mitgliedsstaaten wurden ermutigt, diese freiwillig aus dem Weltnetzwerk zurückzuziehen. Die Herausforderungen bei der Umsetzung der Exit-Strategie waren nicht auf einzelne Länder beschränkt, sondern betrafen alle Mitgliedstaaten mit Biosphärenreservaten aus der Zeit vor Sevilla gleichermaßen. Im Zuge der Qualitätssteigerung haben sich viele UNESCO-Mitgliedstaaten stärker im MAB-Programm engagiert, was in vielen Ländern zu einer Belebung der nationalen Biosphärenreservatsnetzwerke geführt hat.

1 Introduction: The role of UNESCO Biosphere Reserves for global nature conservation and their function as model regions for sustainable development

1.1 Problem statement: Quality enhancement strategies in the Man and the Biosphere Programme to increase the credibility and coherence of the World Network of Biosphere Reserves

> 'Humanity is waging war on nature. We need to rebuild our relationship with it. More than 60 per cent of the world's coral reefs are endangered due to overfishing, destructive practices and climate change. Wildlife populations are plummeting because of overconsumption, population growth and intensive agriculture. And the rate of species extinction is accelerating, with some one million species currently threatened or endangered (...). Deforestation, climate change and the conversion of wilderness for human food production are destroying Earth's web of life. We are part of that fragile web - and we need it to be healthy so we and future generations may thrive. One consequence of our imbalance with nature is the emergence of deadly diseases such as HIV-AIDS, Ebola, and now COVID-19, against which we have little or no defence. Sixty per cent of all known diseases and 75 per cent of new infectious diseases are zoonotic, passing from animals to humans, demonstrating the intimate interconnection between the health of our planet and our own. Biodiversity and ecosystems are essential for human progress and prosperity. They are central to achieving the Sustainable Development Goals and implementing the Paris Agreement on climate change. (...) Let me be clear: degradation of nature is not purely an environmental issue. It spans economics, health, social justice and human rights. (...) By living in harmony with nature, we can avert the worst impacts of climate change and recharge biodiversity for the benefit of people and the planet'. (United Nations Secretary-General Antonio Guterres's remarks to the United Nations Biodiversity Summit, 30 September 2020)

United Nations (UN) Secretary-General Antonio Guterres's strong statement on humanity's war on nature reflects a recognition and awareness of decades of environmental destruction as a serious risk for humanity at even the highest political levels. This perception is linked to a growing concern in society. In global consultations for the 75th anniversary of the UN in 2020, over 1 million people from all countries and walks of life stated that the most pressing medium- and long-term concern was the climate crisis and the destruction of the natural environment (UN, 2020a). Moreover, in the 2020 edition of the World Economic Forum's Global Risks Report, all of the top five long-term risks by likelihood consisted of environmental and climate-related issues for the first time in the history of the survey: extreme weather, climate action failure, natural disaster, biodiversity loss and human-made environmental disasters (World Economic Forum, 2020). Given the overwhelming political, social and economic consensus on the serious long-term consequences for humanity, innovative solutions and effective practical and spatial concepts are urgently needed to combat this global crisis.

Whilst natural ecosystems have become increasingly fragmented for intensive agriculture, industrial development and urban areas (SAUNDERS et al., 1991; GROVE, 1996; HADDAD et al., 2015; WBGU, 2016), protected areas have developed as the most significant and widely accepted strategy for nature conservation (RODRIGUES et al., 2004; ANDAM et al., 2008; GRAY et al., 2016; HOFFMANN et al., 2018). The motivation of setting aside natural areas to maintain their intrinsic values has been part of human history for millennia. From religiously sacred sites to areas for specific resources or species management such as hunting areas set aside for the ruling classes or forest, fish and wildlife reserves. The increasing human pressure on planetary natural resources due to European colonial expansion and commercial enterprise led to the establishment of the first modern protected areas in the second half of the nineteenth century (CHAPE et al., 2005, p. 444). Based on an originally Western paradigm for protected area and often connected to the designation of the Yellowstone National Park in the United States of America in 1872, many protected areas were established to protect spectacular natural features and wildlife in North America, Australia, Europe and South Africa (WATSON et al., 2014, p. 67).

Since the 1960s, the number and total area of protected zones have exponentially grown. As of May 2021, 265,908 protected areas were recorded in the World Database on Protected Areas (WDPA). Most of these areas were on land and covered just over 15% of the earth's land surface (UNEP-WCMC & IUCN, 2021). Currently, Target 11 of the Convention on Biology Diversity (CBD) to conserve 17% of terrestrial and inland water as protected areas by 2020 has not yet been reached (CBD, 2010). In discussions about a post-2020 biodiversity framework, many scientists (DINERSTEIN et al., 2019; WALDRON et al., 2020) and environmentalists (WYSS CAMPAIGN FOR NATURE, 2019; WOODLEY et al., 2021) have called on global leaders to commit to protecting at least 30% of the planet by 2030. So far, negotiations are moving in this direction; in August 2020, an update about the post-2020 global biodiversity framework's zero draft stated under Target 2 to 'protect and conserve through well connected and effective system of protected areas and other effective area-based conservation measures at least 30 per cent of the planet with the focus on areas particularly important for biodiversity' (CBD, 2020, p. 5). The 30% target, which will most likely be adopted at CBD Conference of the Parties 15 in Kunming in May 2022, is a major milestone in protecting the environment, halting the accelerating loss of species and protecting vital ecosystems that are the source of economic security (HAC FOR NATURE AND PEOPLE, 2021). However, a simple increase in the number and total area of protected areas does not automatically yield improvements in conservation outcomes and biodiversity protection (MAXWELL et al., 2020; COAD et al., 2019; GELDMANN et al., 2019). This is because less valuable land is often protected (JOB et al., 2017, p. 1698) and the management quality of protected areas is unknown; thus, some are suspected of being 'paper parks', existing only on paper without management activities (GELDMANN et al., 2015). In a global assessment of over 4,000 protected areas, LEVERINGTON et al. (2010) found that 40% of sites demonstrated major deficiencies and 14% demonstrated significant deficiencies on several management effectiveness indicators; thus, they lacked the basic requirements for effective operation. In addition, PALOMO et al. (2014) noted that many protected areas were threatened due to their spatial isolation and lack of societal support. As a result, the researchers called for a social-ecological approach to protected areas by highlighting their benefits for society, fostering stakeholder involvement and community-based management of protected areas and promoting regional landscape planning beyond the protected areas' limits.

One spatial framework that combines UN Secretary-General Guterres's call for a life in harmony with nature and the socio-ecological approach of greater community participation and regional landscape planning is the United Nations Educational, Scientific, and Cultural Organization (UNESCO) Biosphere Reserves. In the 1970s, UNESCO's Man and the Biosphere (MAB) Programme developed the concept of biosphere reserves to conserve biological diversity, designate areas for ecological and environmental research and provide facilities for education and training. The key mechanism for achieving these objectives was zonation within biosphere reserves, with a strictly protected core zone surrounded by a buffer zone with a decreasing land-use gradient (UNESCO, 1974; BATISSE, 1982). At the time, the concept and vision of biosphere reserves was very innovative. However, the reality of their implementation did not always align with the concept (PRICE, 2002, p. 14). By 1981, the MAB Programme had designated 208 biosphere reserves in 54 countries; however, these 'first-generation' biosphere reserves mainly prioritized the conservation function. They were superimposed on nationally designated protected areas, and the links between environmental resource use and interactions with human settlements for development in the buffer zones were mostly overlooked (BATISSE, 1986; DYER & HOLLAND, 1991; PRICE et al., 2010; COETZER et al., 2014).

The mismatch between concept and reality and the potential link between conservation, human activities and sustainable development were the main topics of the Second International Biosphere Reserve Congress in 1995 in Seville (PRICE et al., 2010). The Seville conference led to significant changes in the MAB Programme, transforming the initial concept of a research programme into a modern instrument for the dual mandate of conservation and sustainable development (KÖCK & ARNBERGER, 2017). Local people and their activities were explicitly integrated into MAB, and reflected in the revision of the biosphere reserves' spatial zoning. In addition to the core and buffer zones, a third zone called the transition zone was introduced. It included human settlements and allowed for cooperation between different actors and sustainable socio-economic development activities. Unlike the core zone, the buffer and transition zones did not need to be legally protected areas (JOB et al., 2017). Another important outcome of the Seville Strategy was the adoption of the Statutory Framework for the World Network of Biosphere Reserves, which defined the principles, criteria and designation procedure for biosphere reserves (BRIDGEWATER, 2016). Article 9 states that

> 'the status of each biosphere reserve should be subject to a periodic review every ten years, based on a report prepared by the concerned authority, on the basis of the criteria of Article 4, and forwarded to the secretariat by the State concerned. The report will be considered by the Advisory Committee for Biosphere Reserves for recommendation to the ICC'. (UNESCO, 1996, p. 18)

Over the last 10 years, the MAB Programme has increasingly focused on improving quality of biosphere reserves through a periodic review process to ensure that all biosphere reserves in the World Network of Biosphere Reserves conform to current requirements (JOB et al., 2017; KÖCK et al., 2020, chapter 3.3). Whilst many governments have submitted an increasing number of periodic reviews over the years, it became clear that a large proportion of biosphere reserves designated before 1996 did not include all three zones (ISHWARAN, 2012; COETZER et al., 2014; PRICE, 2017). In 2013, the International Coordination Council of the MAB Programme (MAB ICC) announced the so-called 'Exit Strategy', as 'a three-stage process to manage the periodic review process as a tool to assess, monitor and improve the quality of the World Network of Biosphere Reserves' (UNESCO, 2013, p. 39). All countries are required to submit periodic reviews to show that all their biosphere reserves conform to the criteria (UNESCO, 2013). Whilst this decision within the UN system was a firm step towards quality improvement within the World Network of Biosphere Reserves (KÖCK & ARNBERGER, 2017, p. 87), it also meant that several member states were faced with the possibility of exclusion from the World Network of Biosphere Reserves. At the time, 266 biosphere reserves in 76 countries were affected by the Exit Strategy (UNESCO, 2014, p. 29). In 2017, one year after the envisioned deadline for completing the Exit Strategy, 126 sites met the criteria, whilst 25 sites announced withdrawal from the World Network of Biosphere Reserves. As a result, the MAB ICC decided to extend the Exit Strategy until 2020 and institute the Process of Excellence and Enhancement of the World Network of Biosphere Reserves as well as quality improvement of all members of the World Network to ensure that all biosphere reserves served as models for the implementation of the 2030 Agenda and the associated Sustainable Development Goals (SDGs) (UNESCO, 2017a). In September 2020, the MAB Secretariat announced that significant progress had been made since the Process of Excellence and Enhancement was established in 2017. Seven countries decided to voluntarily withdraw an additional 21 biosphere reserves from the World Network of Biosphere Reserves, and three sites were excluded from the process due to being in situations of conflict. However, 45 biosphere reserves remained part of the Process of Excellence and Enhancement (UNESCO, 2020a). Due to the COVID-19 pandemic, the process formerly known as the Exit Strategy and now known as the Process of Excellence and Enhancement was completed at the MAB ICC meeting in Abuja, Nigeria in September 2021 (UNESCO, 2022).

Reporting processes are not only important for biosphere reserves but also most other protected areas and international designations. Even in the UNESCO system, biosphere reserves (714 sites) are only one of three approaches and designations for nature conservation, the other two being the famous natural UNESCO World Heritage Sites (213 sites) and since 2015 the UNESCO Global Geoparks (140 sites). These UNESCO designations sometimes overlap and create Multi Internationally Designated Areas (MIDAs; SCHAAF & CLAMOTE RODRIGUES, 2016). JOB et al. (2017) identified several benefits of multi-layered management approaches such as the division into different management zones, stronger monitoring standards and economically competitive local brands. However, other scientists are concerned that the multitude of designations complicates the management and evaluation of sites due to the multiple layers of governance and institutional requirements (MATAR & ANTHONY, 2017), which is also reflected in additional reporting processes (PRICE et al., 2010). So far, these monitoring processes still lack harmonization and information sharing. The World Heritage Sites, for example, encompass of a six-year evaluation procedure, the Global Geoparks are subject to a periodic review every four years and the Biosphere Reserves have a 10-year periodic review system (PRICE et al., 2010).

1.2 Objectives and research questions

> 'The elegance of the biosphere reserve concept lies in its simplicity; yet the practice of converting the concept's implications into reality at international, national and local scales raises a number of challenges'. (ISHWARAN et al., 2008, p. 119)

This thesis concerns one of the greatest challenges within the global UNESCO Biosphere Reserves: how to respect cultural, social and economic diversity and the national sovereignty of each UNESCO member state whilst ensuring a uniform framework to guarantee a certain level of credibility and coherence. Although concept of biosphere reserves became dormant after its development in the 1970s, it is now well-suited to achieving the 2030 Agenda 2030 and the SDGs. To live up to their reputation as model regions for sustainable development, the biosphere reserves have been resurrected in recent years with an internal quality improvement process. This thesis presents a global assessment of the challenges associated with the recently introduced Exit Strategy and Process of Excellence and Enhancement. Therefore, the underlying guiding research questions are as follows: 'Does the current UNESCO World Network of Biosphere Reserves fulfil the quality requirements of the Statutory Framework as model regions for sustainable development?' and 'How do they relate to the other UNESCO labels?'

The objectives of this thesis are twofold. First, it attempts to construct a general understanding of the diversity of the World Network of Biosphere Reserves through the presentation of 20 country-specific case studies, which showcase challenges in implementing the biosphere reserve concept in general and the periodic review process in particular. Second, the thesis abstracts and presents information gained from expert interviews to identify the current greatest challenges in the World Network of Biosphere Reserves, differences in quality management between member states, the effects of recently introduced quality enhancement strategies and the interlinkages with other UNESCO designations. The selected experts contributed both their country-specific experiences with the national implementation of the biosphere reserves and international experiences and expectations with MAB governance structures. Finally, this thesis provides suggestions for future actions within the MAB Programme.

Given these objectives, the thesis attempts to answer the following research questions:

- What are differences in quality management and the periodic review process of biosphere reserves between member states?
- What effects did the Exit Strategy and later the Process of Excellence and Enhancement have on the MAB Programme in general and the 76 directly affected member states and their 266 biosphere reserves in particular?
- What kind of new ways and methods could be used to improve the evaluation process?
- What are the interlinkages between the three UNESCO designations: World Heritage Sites, Biosphere Reserves and Global Geoparks? Is it a complementary system of labels, or do they compete against each other?

1.3 Thesis structure

The historical development of the nature conservation approach and the globalization of protected areas is covered in detail in the second chapter of this thesis, which offers an initial theoretical and contextual approach to the research topic. Chapter 3 addresses the transition from classic protected areas to the concept of biosphere reserves. It also presents the historical background for the drastic paradigm shift in the concept of biosphere reserves since the Seville Conference. Then, the chapter provides a detailed explanation of periodic reviews and introduces of the Exit Strategy as the MAB Programme's quality enhancement approach. It concludes with a description of the World Network of Biosphere Reserves' spatial diffusion, presents the most important actors and structures in the network and provides an initial overview of interlinkages with the other UNESCO labels for nature conservation based on the literature.

In the fourth chapter, the methodological approach used for the research is presented, and an overview of qualitative methods, research materials and data collection is provided. The chapter also describes the selection of experts and creation of the semi-structured interview questionnaire. In addition, it presents the qualitative content analysis process used to answer the research questions.

The fifth chapter presents the initial findings from the study using 20 country-specific case studies. The aim is to illustrate the complexity of and challenges in the national implementation of the biosphere reserve concept and quality management in an intergovernmental programme throughout the different UN regions. In each case study, a map of the national biosphere reserves' natural area classifications is provided, in addition to basic information about the biosphere reserves and national structures used to manage them. Finally, the greatest current challenges for each country and future plans are presented based on information from the expert interviews.

The expert interviews also form the basis for the data analysis in Chapter 6; the collected data is abstracted in relation to the research questions. The main challenges faced by participants from the World Network of Biosphere Reserves are covered first, followed by the introduction and implementation of quality enhancement strategies, possible improvements to the periodic review process and interlinkages with other UNESCO labels, particularly the World Heritage Sites and Global Geoparks. The chapter concludes by addressing the question of the interviewed experts of how many UNESCO Biosphere Reserves the world network will need in the future. Subsequently, the results from the empirical analyses are synthesized in Chapter 7, which presents an overall picture of international challenges and differences in implementation. In addition, the research questions are extensively discussed in light of the findings. Finally, concluding remarks for the thesis and the research are offered in Chapter 8.

2 Nature conservation and protected areas

2.1 Definition of key terms and historical overview

The word and the concept of nature is complex, because of its different meanings and usages not only in everyday parlance but also in science. In the latter field, nature refers to the relationships between different disciplines. Sometimes, nature is even seen as the only study of all science.

The word 'nature' derives from the Latin word 'natura' meaning the course of things, the original natural character and in ancient times meant literally 'birth' originated from 'natus' (HARPER, n.d.). As early as the mid-14th century, nature was described 'as the forces or processes of the material world; that which produces living things and maintains order' (HARPER, n.d.). This original and creational view of nature narrowed with population growth in the 1660s and came to be understood as 'the material world beyond human civilization or society; an original, wild, undomesticated condition' (HARPER, n.d.). It is notable that nature was described as early as the 17th century as 'wild' and 'undomesticated', two terms that are at the heart of a fundamental debate on the conservation of nature in light of exponential population growth in the 21st century.

According to CASTREE (2014), nature currently has four main meanings (see Figure 1). First, it refers to the nonhuman world, especially all areas that are untouched, pristine or barely affected by humans (external nature). Second, it refers to the physical world in its entirety, including humans as biological entities and products of evolutionary history (universal nature). Third, it refers to the essential quality or defining features of a thing; it is natural for birds to fly and fish to swim (intrinsic nature). Finally, it refers to the power or force that governs some or all living things such as gravity (superordinate nature; p. 10).

Figure 1: The four main meanings of nature

NATURE			
The non-human world of living and inanimate phenomena, be they 'pristine' or modified.	The physical world in its entirety, including human beings as both products of natural history and present-day biological organisms.	The defining features or distinguishing quality of living and inanimate phenomena, including human beings.	The power, force or organising principle animating living phenomena and operating in or on inanimate phenomena.
'EXTERNAL NATURE'	'UNIVERSAL NATURE'	'INTRINSIC NATURE'	'SUPER-ORDINATE NATURE'

Source: CASTREE, 2014, p. 10

Although it is clear that the word 'nature' is used in a wide range of mundane and specialized situations, the concept of nature is not exclusively associated with the word. Instead, its meanings are signified by a range of other words that are or have become part of the collective vocabulary. In the 21st century, supplementary terms such as 'environment', 'wilderness', 'biodiversity' and 'ecosystem' arose and are nowadays largely used synonymously. Some of them encompass both human and nonhuman nature, which depends on the context in which they are invoked (CASTREE, 2017). The word 'biodiversity' originated in the 1980s as a contraction of 'biological diversity' and coincided with the emergence of two new scientific fields: genetic engineering and conservation biology. Biodiversity refers to diversity within the entirety of the world's biosphere which encompasses genetic, species and habitat diversity (NEUMANN, 2009, p. 308).

'Conservation' is another broad term that can refer to multiple phenomena, such as the conservation of art, buildings, places and heritages sites but also of matter and energy. However, conservation of nature refers to 'norms, values, institutions and social movements concerned with the protection or wise use of natural areas and living resources' (JAY & MORAD, 2009, p. 259). The motivation to protect natural areas and living resources to maintain their intrinsic value, has been a part of human history for millennia (CHAPE et al., 2005).

Whilst pollen analysis indicates that the deforestation of temperate forests began in Mesolithic and Neolithic times to facilitate agriculture and provide charcoal or wood for construction, it is more difficult to assess the extent to which a distinctive conservationist response developed. However, written literature shows that, as early as the fourth millennium BC, some advocates were aware of the destructive power of early agrarian and hydraulic empires and the likely consequences of uncontrolled deforestation. Increasing deforestation and wood shortages due to expanding military activities, shipbuilding and state building were also evident during the classical Greek and Roman era (GROVE, 1996).

During the pre-Islamic period, traditional forms of protected areas or resource reserves, such as the *hima*, were established, and an ancient acknowledgement of the scarcity of renewable resources was evident. The Koran attaches great importance to the value of preserving one's natural heritage, and the concept of the *hima* was given legal standing in *shari'ah* law. In addition, the Prophet Muhammed established a legal system to govern these protected areas, which provide communal benefits (SULAYEM et al., 1994, p. 79). The first documented reference to nature conservation was an edict from Emperor Asoka of India in 252 BC for the protection of animals, fish and forests. The practice of establishing sacred areas as religious sanctuaries or exclusive hunting reserves was very common and continued throughout the region until the present (MISHRA, 1994, p. 181). In East Asia, the Japanese emperor declared a bird hunting and preservation section in the Imperial government during the seventh century AD (XIANPU, 1994, p. 161). In Latin America, the Mayans already incorporated management and protection areas for extraction and untouchable reserves. At the time of the Spanish Conquest, the Aztec emperor Montezuma maintained several protected areas and a zoological and botanical garden (MILNE & WAUGH, 1994, p. 281).

In the Middle Ages, the eastward movement of populations in Europe was closely associated with forest clearance. Awareness of the extent of deforestation, particularly in Germany, led to notions of environmental control. Local regulations against deforestation to protect the timber supply were developed in Nürnberg in southern Germany as early as 1309 under the Nürnberg Ordinance, which required the restoration of forests that had been cleared in the preceding 50 years and transformed into cultivated fields (GROVE, 1996). A new and much more complex way of viewing the relationship between people and nature arose with European colonial expansion and commercial enterprise. A sophisticated awareness of the growing capacity of people to alter their physical surroundings, the 'search for

Eden' and experiences of encountering new lands, peoples, animals and plants helped promote a new social significance attached to nature (GROVE, 1996; CHAPE et al., 2005).

The modern understanding of nature conservation is largely a Western concept that dates from the late 19th and 20th centuries. It is associated with the creation and management of protected natural areas, such as national parks, game reserves and wildlife sanctuaries, due to increasing human pressure on planetary natural resources (JAY & MORAD, 2009). As tools for nature conservation, the establishment of protected areas reflects the conceptual division between nature and human society and has profound political significance, as the state makes rules about who can use nature and where, when and how it can be used. This way of thinking has deep roots in Western societies and therefore must be understood in the historical context of the wider political structures of colonial societies and neo-colonial relationships between the countries of the West and the Global South (ADAMS & HUTTON, 2007).

2.2 National Parks in the United States of America and Game Reserves in the British Empire

JAY and MORAD (2009) identified two main roots for the understanding of modern nature conservation: (1) an elitist tradition that originates from the British association between aristocracy and hunting and game management and (2) a populist tradition that evolved from romantic trends in American society that linked wilderness to moral virtue and national identity. When researching nature conservation on a global scale, it is important to distinguish between two different views of nature conservation: that of colonialists and the people of Europe during the 19th century. The colonialists perceived the lands of North America, New Zealand, Australia and Africa as largely empty. They set aside areas protected from human encroachment and thus created landscapes that were clearly divided between human habitation and wilderness. By contrast, countries that were already densely populated at the time, such as in Europe, were more concerned with conserving traditional landscapes that had been constructed and managed over generations against the encroachment of modern development, incorporating cultural and natural heritage in nature conservation (pp. 260, 262-263).

In the United States of America, concerns over nature conservation began in the 19th century. In 1836, the writer and philosopher Ralph Waldo Emerson articulated the importance of nature as a source of moral and spiritual uplift: 'In the woods, we return to reason and faith. There I feel that nothing can befall me in life, no disgrace, no calamity (…) which nature cannot repair' (EMERSON, 1836, p. 4). The conquest of the land west of the Mississippi River in the 1850s and 1860s revealed new and unparalleled landscapes. Travel books and romantic stories of adventures, explorers and artists linked romantic beauty and heroic grandeur with wild and empty spaces. As a result, public support for preserving areas of nature and wilderness steadily grew (JAY & MORAD, 2009, p. 260). A senator from California, John Conness, introduced a bill to the U.S. Senate in 1864 to authorize for the first time in history a federal grant to the State of California for the Yosemite Valley and the Mariposa Big Tree Grove (YOSEMITE ACT, 1864, p. 203). The Yosemite Bill was signed by

President Abraham Lincoln in 1864 and was widely supported across the United States of America. A similar development occurred around the geysers, lakes and waterfalls of Yellowstone in Wyoming, which were notable natural features. After sustained lobbying, the Yellowstone Park Act was passed in March 1872 and stated that

> 'the tract of land in the Territories of Montana and Wyoming, lying near the headwaters of the Yellowstone River (...) is hereby reserved and withdrawn from settlement, occupancy, or sale under the laws of the United States, and dedicated and set apart as a public park or pleasuring-ground for the benefit and enjoyment of the people'. (YELLOWSTONE NATIONAL PARK PROTECTION ACT, 1872)

This led to the creation of the world's first national park. The American concept of national parks was soon adapted in Canada (Banff National Park, 1887), Australia (Royal National Park, 1879 and Belair National Park 1891) and New Zealand (Tongariro National Park, 1894; ADAMS 2004, p. 78).

In 1892, John Muir founded the Sierra Club, the oldest nature conservation organization in the United States of America. Nature conservation received a boost from President Theodore Roosevelt (1901–1909) who strongly believed in the moral benefits of contact with wilderness and recreation. He emphasized that the land itself embodied the ideals of American independence, resourcefulness, and democratic freedom (JAY & MORAD, 2009, pp. 260-261). These views were founded on a conception of nature as pristine. Indeed, wilderness remains an important element of national identity in the United States of America (ADAMS & HUTTON, 2007, p. 153).

Early US-American national parks were a cornerstone of conservation around the world. According to ADAMS (2004), they had three characteristic features that would have implications for future generations of conservationists around the world. First, they were created based on the alleged premise that the lands were completely natural, unmanaged and uninhabited by humans. This belief in a lack of human presence in remote and natural-seeming areas was widely translated to other regions, particularly Australia, Sub-Saharan Africa or the Amazon (AMEND & AMEND, 1995; POIRIER & OSTERGREN, 2002; BROCKINGTON & IGOE, 2006; NAUGHTON-TREVES et al., 2006; MILGROOM & SPIERENBURG, 2008; SHEPARD JR. et al., 2010). Second, the national parks represented a militaristic approach to management. After years of ineffectual civil administration of Yellowstone National Park, the U.S. Army took control in 1886 and remained for 32 years. When the National Park Service took over management of the park in 1918, they copied the army's uniforms and hired former soldiers as park rangers. Third, national parks developed a close relationship with tourism. Their advocates chose to promote tourism at national parks in the hopes of demonstrating their value, both in terms of economic benefits and wider public acceptance (ADAMS, 2004, p. 79; WELLS, 1996).

In the United Kingdom, early conservation practices were largely related to the protection of habitat for game birds, deer, trout, and other species hunted by aristocrats. Whilst concern over nature conservation was widespread in society at the beginning of the 19th century, entrenched interests such as the owners of large estates, competing government departments and local county councils resisted the implementation and designation of protected areas. It was only after the end of the Second World War that a National Park Commission was established in 1949. However, voluntary organizations in the United Kingdom had a major influence on general public opinion and nature conservation in other parts of the world (JAY & MORAD, 2009, pp. 261-262).

In the British Empire, the idea of special areas for game and wildlife conservation was more easily realized than in the United Kingdom. Under British colonial rule, African territories were strongly influenced by the British tradition of game protection for the elite in form of resorts for gentlemen hunters, for travellers or colonial servants. In the colonial discourse, these game reserves were idealized as wild and exotic lands where colonial youth could develop a sporting spirit. Game reserves focused on the protection of game animals and their habitats and were largely a response to uncontrolled hunting by early explorers and settlers (JAY & MORAD, 2009, p. 263; ADAMS, 2004, p. 68). Many scientists, including NEUMANN (1996) and GROVE (1996), argued that, to some Europeans, Africa represented a lost Eden in need of protection. However, their commitment was motivated not only by their efforts to protect an Edenic Africa but also the loss of their own privileged lives in rural Great Britain. For many conservationists at the time, the disappearance of an idealized British countryside was often as important as the disappearance of an idealized African wild. By the beginning of the 20th century, game reserves had been established in Sudan, Kenya, Uganda and the British colonies of central Africa (JAY & MORAD, 2009, p. 263).

In 20th century, the experiences of national parks in the United States of America, Canada, Australia and New Zealand gained growing interest amongst colonial conservationists for several important reasons. First, they required much greater commitment from governments with (more) legally binding borders than other protected area categories, powerful governing boards and proper funding, as any failure would reflect badly on the nation's reputation. Second, they were declared to be in the interest of the entire country. Third, the national parks model could generate much more substantial income from tourism than game reserves. These economic benefits were particularly relevant during the economic recession of the late 1920 and 1930s (ADAMS, 2004, pp. 88-89). Consequently, most national parks and game reserves in the colonial time were designated between the two world wars (OLINDO & MBAELELE, 1994, p. 49). The first national park in Africa was established in the Belgian Congo in April 1925. The Albert National Park, named after the king of Belgium at the time, encompassed the habitat of the charismatic mountain gorillas. Later, it was extended and renamed Virunga National Park (VERWEIJEN & MARIJNEN 2018, pp. 10-11). In South Africa, the Sabi and Singwitsi game reserves were merged and declared a national park in 1926: the now famous Kruger National Park. This occurred at a time when the idea that viewing and studying wildlife was a legitimate and economically viable form of land use and that state land and finance should be allocated to this purpose gained increasing acceptance amongst white South Africans. At the same time, an Afrikaner nationalism emerged in the search for a white South African national identity. The nationalistic interpretation of wildlife conservation bears similarities with the creation of national parks in the United States of America and Australia (MABUNDA et al., 2003; CARRUTHERS, 1995).

In French West Africa, a decree in 1935 regulated hunting and established national parks as refuges for certain animal species. The establishment of protected areas was generally conducted in places with a low population density to create buffer zones between people in areas of local conflict and for ecological reasons, such as minimizing diseases (e.g., sleeping sickness and onchocerciasis), preventing natural soil depletion and addressing lack of water resources. Thus, these areas were rich with wildlife and ideal sites for protection. Mainly established during the 1930s and 1950s, the most important areas were in eastern Senegal, southwest Mali and southeast and southwest Burkina Faso (OLINDO & MBAELELE 1994).

In Tanzania, the Serengeti was made into a game reserve in 1908 to protect its lions. After several years of dispute with the Maasai, Sukuma and Ndorobo people, it was officially recognized as national park in 1951. In 1957, the director of Frankfurt Zoo and the Frankfurt Zoological Society, Bernhard Grzimek, and his son Michael flew to the Serengeti to undertake an aerial survey of plain animals and their big migrations. In his best-selling book, he wrote:

> 'Large cities continue to proliferate. In the coming decades and centuries, men will not travel to view marvels of engineering, but they will leave the dusty towns in order to behold the last places on earth where God's creatures are peacefully living. Countries which have preserved such places will be envied by other nations and visited by streams of tourists. There is a difference between wild animals living a natural life and famous buildings. Palaces can be rebuilt if they are destroyed in wartime, but once the wild animals of the Serengeti are exterminated no power on earth can bring them back.' (GRZIMEK 1959)

His vision was disseminated across elite networks of likeminded people between the Americas and Europe, then from colonial empires to the world. It reflected a nature conservationist aim for the protection of particular places over the entirety of the 20th century (ADAMS, 2004, p. 69).

The model of nature conservation in this century was based mainly on the U.S. idea of a national park as a pristine and wilderness area, and the British concept of managed game reserves. In sub-Saharan Africa in particular, this approach of creating protected areas, excluding people as inhabitants of the areas and rejecting human use of natural resources within the protected areas has been called 'fortress conservation' or the 'fences and fines approach' (HUTTON et al., 2005; WELLS, 1996; ANDERSON & GROVE, 1987; NEUMANN, 1998; BROCKINGTON & SCHMIDT-SOLTAU, 2004).

2.3 Globalization of nature conservation after the Second World War

Since the Second World War, the general rise of international institutions is also observable in nature conservation. In 1948, the first truly international organization was established: the well-known International Union for the Conservation of Nature (IUCN). As a membership union, it consists of both government and civil society organizations and is headquartered in Gland, Switzerland (JAY & MORAD, 2009, p. 263; IUCN, 2019). As of November 2017, over 85 states, 120 government agencies, 1,000 non-governmental organizations and 15 indigenous people's organizations were member of IUCN. Since its founding, IUCN members have convened every two to four years to debate major policy issues and promote the global conservation community's priorities (IUCN, 2018).

Since the 1960s, the number and total area of protected areas has dramatically grown. As of May 2021, there are 265,908 protected areas around the world (UNEP-WCMC &

IUCN, 2021). According to JAY and MORAD (2009), this rise in global nature conservation can be divided into two main stages according to philosophy and development. Spanning the early 1960s to the 1980s, the first stage was characterized as an age of realization about the global scale of environmental issues and their interconnectedness. The consequences of industrial development, such as acid rain, caused damage in Scandinavia and eastern Canada, and the clear link between the decline of bird populations and the application of pesticides resulted in widespread awareness of the interconnectedness of life (pp. 263-265). This led to several important publications, events and agreements in the 1970s. The most influential one was the UN Conference on the Human Environment in Stockholm in 1972, the first international conference to tackle environmental and development questions on a global scale. In total, representatives from 113 countries and 400 intergovernmental and non-governmental organizations gathered to discuss environmental issues, differences between countries and the importance of development for poor countries (JAY & MORAD, 2009, p. 265; NORMAN & CARR, 2009, pp. 406-407). This conference also led to the foundation of the UN Environment Programme (UNEP) in 1973 to coordinate the development of environmental policy by monitoring the global environment and bringing emerging issues to the attention of governments and the international community. Furthermore, the MAB Programme was launched by UNESCO in 1968 and the first biosphere reserves were established in 1976 (see Chapter 3). The Ramsar Convention was signed in 1971 to ensure the conservation and wise use of wetlands, the Paris Convention for the Protection of the World Cultural and Natural Heritage was ratified in 1972 to establish a system of World Heritage Sites (WHS) as cultural or natural features of outstanding universal value and the Convention on International Trade in Endangered Species of Wild Fauna and Flora was adopted in 1973 to ensure that international trade in wild animals and plants did not threaten their survival. Finally, in 1980, the World Conservation Strategy was published by IUCN, UNEP and the World Wide Fund for Nature (WWF). It recognized that nature conservation was inseparable from social and economic sustainability and could not be achieved without development to alleviate poverty. It was the first major public document to introduce the term 'sustainable development' (JAY & MORAD, 2009, p. 265; JOB et al., 2017, p. 1701).

Spanning the beginning of the 1980s to the present, the second stage is characterised by the gradual acceptance in society that nature conservation is dependent on the preservation of ecosystems, ecosystem processes and interactions between, plants, animals and humans and that successful nature conservation requires the acceptance and involvement of local populations. Due to rapid growth of protected areas during the second half of the 20th century, contact and conflicts between protected areas and local communities also steadily increased comparted to previous centuries. The role that protected areas can play in poverty alleviation and economic development in surrounding communities was also increasingly acknowledged (WATSON et al., 2014, pp. 67-68). A notable finding was that national and international efforts and policies will not have any impact if local people cannot or will not create conditions for the protection of threatened species. This resulted in a paradigm shift from previous species protection (e.g., tiger, big game) through governmental decrees to more far-reaching nature conservation programmes with the participation of local communities, also to support sustaining their livelihoods (JAY & MORAD, 2009, p. 264).

This development is very closely connected to the United Nations Conference on Environment and Development (widely known as the Rio Earth Summit) in June 1992 in Rio de Janeiro. The conference was the result of the UN General Assembly's call for a global meeting to address the issues raised in the 1987 Brundtland report *Our Common Future,* which decried the world's failure to achieve sustainable development and outlined several actions needed to 'ensure both sustainable human progress and human survival' (UN, 1987, p. 11). The Rio summit focused on developing a global framework to address environmental degradation through sustainable development. The main results and agendas were condensed into several documents, such as the Agenda 21, the Rio Declaration on Environment and Development and the most influential conventions to date: the UN Framework Convention on Climate Change and the UN CBD (NORMAN & CARR, 2009, pp. 406-407). This emergence of international conservation targets and conventions in the 1990s marked a new era in the globalization of conservation activity (ANTHAMATTEN & HAZEN, 2015; ZIMMERER et al., 2004).

At the Millennium Summit in 2000, all 191 UN member states established eight Millennium Development Goals (MDGs) to be met by 2015. Goal 7 aimed to ensure environmental sustainability through four targets, two of which tackled nature conservation. Target 7.A aimed to 'integrate the principles of sustainable development into country policies and programmes and reverse the loss of environmental resources', whilst Target 7.B aimed to 'reduce biodiversity loss, achieving, by 2010, a significant reduction in the rate of loss' (UN, 2000). In 2005, the Millennium Ecosystem Assessment, a report by 1,300 experts from 95 countries, concluded that approximately 60% of the earth's ecosystem services were being degraded or unsustainably used (MILLENIUM ECOSYSTEM ASSESSMENT, 2005, p. 1). In 2015, the final analysis of the MDGs concluded that, with regard to Targets 7.A and 7.B, deforestation had slowed but continued to jeopardize species and the livelihoods of millions of people. In many countries, deforestation remained alarmingly high. South America and Africa experienced the largest net losses of forest area in the first 10 years of the 21st century (UN, 2015a, p. 52). Although Latin America and the Caribbean, Oceania and Western Asia substantially increased their terrestrial protected areas since 1990, there is still in a race against time to save animals and plants from extinction (UN, 2015a, pp. 56-57).

Over the past 10 years two major strategies demonstrated the rise in nature conservation efforts and the importance of conserving nature, biodiversity and ecosystems. First, in 2010, the parties to the CBD adopted the Strategic Plan for Biodiversity 2010-2020 and 20 Aichi Biodiversity Targets. The strategy has since been endorsed by several multilateral environmental agreements as a global framework (UNEP-WCMC & IUCN, 2016, p. V). Specifically, Aichi Biodiversity Target 11 calls for the following:

> 'By 2020, at least 17 per cent of terrestrial and inland water areas and 10 per cent of coastal and marine areas, especially areas of particular importance for biodiversity and ecosystem services, are conserved through effectively and equitably managed, ecologically representative and well-connected systems of protected areas and other effective area-based conservation measures, and integrated into the wider landscape and seascape'. (CBD, 2010, p. 9)

In 2015, UN member states adopted the second major global strategy, the 2030 Agenda for Sustainable Development, and its 17 SDGs, which replaced the MDGs. Both agreements represent the most important environment and sustainable development commitments ever made by governments in the international arena. They recognise the important role of protected areas as a key strategy for biodiversity conservation and sustainable development (UNEP-WCMC & IUCN, 2016, p. V). SDG 14 aims to 'conserve and sustainably use the oceans, seas and marine resources for sustainable development', whilst SDG 15 aims to 'protect, restore and promote sustainable use of terrestrial ecosystems, sustainably manage forests, combat desertification, and halt and reverse land degradation and halt biodiversity loss' (UN, 2015b, p. 18).

As of October 2020, Aichi Biodiversity Target 11 has not yet to be achieved. Currently around 15% of the earth's land surface is designated as protected areas (UNEP-WCMC et al., 2020). In the debate for a post-2020 biodiversity framework, many scientists (DINERSTEIN et al., 2019; WALDRON et al., 2020) and environmentalists (WYSS CAMPAIGN FOR NATURE, 2019; WOODLEY et al., 2021) have called on global leaders to commit to protecting at least 30% of the planet by 2030. So far, negotiations are moving in this direction; in August 2020, the update of the post-2020 global biodiversity framework's zero draft stated that Target 2 would 'protect and conserve through well connected and effective system of protected areas and other effective area-based conservation measures at least 30% of the planet with the focus on areas particularly important for biodiversity' (CBD, 2020, p. 5). The 30% target would be a major milestone in protecting the environment and preventing further fragmentation of ecosystems. However, doubling the global area of protected areas in only 10 years is an ambitious goal; it would undoubtedly provoke, increase and reinforce debates on the performance (WATSON et al., 2014), conservation outcomes (OLDEKOP et al., 2016) and management effectiveness (LE SAOUT et al., 2013; GELDMANN et al., 2015; COAD et al., 2019) of designated protected areas.

2.4 Protected areas by IUCN definition and global coverage

The intertwining of conservation and globalization efforts in recent years has led to an unprecedented variety and scope of spatial conservation frameworks whose purposes, management goals and activities may vary from strict nature protection to sustainable utilization of natural resources. The global increase in conservation areas has developed on basis of existing protected area frameworks, such as the aforementioned national parks and game reserves (where human access is restricted), and the rapid evolution of new management areas, such as community conservation areas, watershed-based projects, and buffer zones in biosphere reserves, where conservation is integrated into traditional human lifestyles or even takes place alongside sustainable resource extraction (ZIMMERER, 2006, pp. 66-67; DUDLEY, 2008, p. 3). This variety reflects the fact that conservation is not achieved in the same manner in every situation; what may be desirable or feasible in one place could be counter-productive or politically impossible in another. To provide guidance on different approaches, IUCN agreed on a definition of a protected area and identified six categories of protected areas based on management objectives (DUDLEY, 2008, p. 3). Since 2008, IUCN has defined a protected area as 'a clearly defined geographical space, recognised, dedicated and managed, through legal or other effective means, to achieve the long-term conservation of nature with associated ecosystem services and cultural values' (Dudley, 2008, p. 8).

The six categories of protected areas are organized in decreasing order of use and human influence. They range from Category I (strict nature reserves and wilderness areas) to Category VI (protected areas with sustainable use of natural resources; see Table 1). These categories are useful as a global standard for defining, comparing and communicating about protected areas, as they offer a common language for describing protected areas on a global scale. They form the basis for the WDPA maintained by IUCN and the UNEP World Conservation Monitoring Centre (UNEP-WCMC; Borrini-Feyerabend et al., 2013, p. 8).

Table 1: IUCN protected area categories

Protected area category		Management objectives
Ia	Strict nature reserve	Strictly protected for biodiversity and possibly geological or geomorphological features, where human visitation, use and impacts are controlled and limited to ensure protection of the conservation values
Ib	Wilderness area	Usually large unmodified or slightly modified areas, retaining their natural character and influence, without permanent or significant human habitation, protected and managed to preserve their natural condition
II	National park	Large natural or near-natural areas protecting large-scale ecological processes with characteristic species and ecosystems, which also have environmentally and culturally compatible spiritual, scientific, educational, recreational and visitor opportunities
III	Natural monument or feature	Areas set aside to protect a specific natural monument, which can be a landform, sea mount, marine cavern, geological feature such as a cave, or a living feature such as an ancient grove
IV	Habitat/species management area	Areas to protect particular species or habitats, where management reflects this priority. Many will need regular, active interventions to meet the needs of particular species or habitats, but this is not a requirement of the category
V	Protected landscape or seascape	Where the interaction of people and nature over time has produced a distinct character with significant ecological, biological, cultural and scenic value and where safeguarding the integrity of this interaction is vital to protecting and sustaining the area and its associated nature conservation and other values
VI	Protected areas with sustainable use of natural resources	Areas which conserve ecosystems, together with associated cultural values and traditional natural resource management systems. Generally large, mainly in a natural condition, with a proportion under sustainable natural resource management and where low-level non-industrial natural resource use compatible with nature conservation is seen as one of the main aims

Source: Borrini-Feyerabend et al., 2013, p. 9; Leung et al., 2018, p. i

However, each country has its own legislation to designate protected areas. Thus, there are hundreds of different types and names for protected areas. In some cases, larger landscapes are a complex of different protected area categories. Furthermore, international institutions such as UNESCO have created further designations to protect key habitats such as Biosphere Reserves, World Heritage Sites, Ramsar sites, and Global Geoparks (BORRINI-FEYERABEND et al., 2013, p. 8). These international designations were originally identified as categories of their own. The first category system for protected areas was developed by IUCN in 1978 and included 10 categories divided into three groups. Whilst all protected areas in Group A and B were mainly directly related to IUCN, those in Group C were part of international programmes. Category IX encompassed biosphere reserves, and Category X encompassed natural World Heritage Sites (IUCN, 1978, pp. 10-11). However, over time, several limitations of this system became apparent. In particular, the fact that the 10 categories of protected could not always be exclusively considered caused confusion (PHILIPPS, 2004, p. 8). A task force was established to review the system and recommended that a new system be developed around Categories I-V from 1978, abandoning Categories VI-X. Consequently, the international designations were excluded from the standard categories in the revised version of the IUCN guidelines in 1994 (DUDLEY, 2008, p. 4).

Confusion and uncertainty about the relationship between biosphere reserves and the IUCN protected area management categories was rooted in the fact that the concept of biosphere reserves embodied a range of management objectives in different spatial zones, each of which corresponded to a different category in the IUCN system (see Chapter 3). In addition, parts of some biosphere reserves may not fit the criteria of a protected area at all, as the integrative approach in a broader landscape concept was the main value of biosphere reserves (BRIDGEWATER et al., 1996, pp. 8-9).

Figure 2: Growth in total number of protected areas from 1930 to 2020

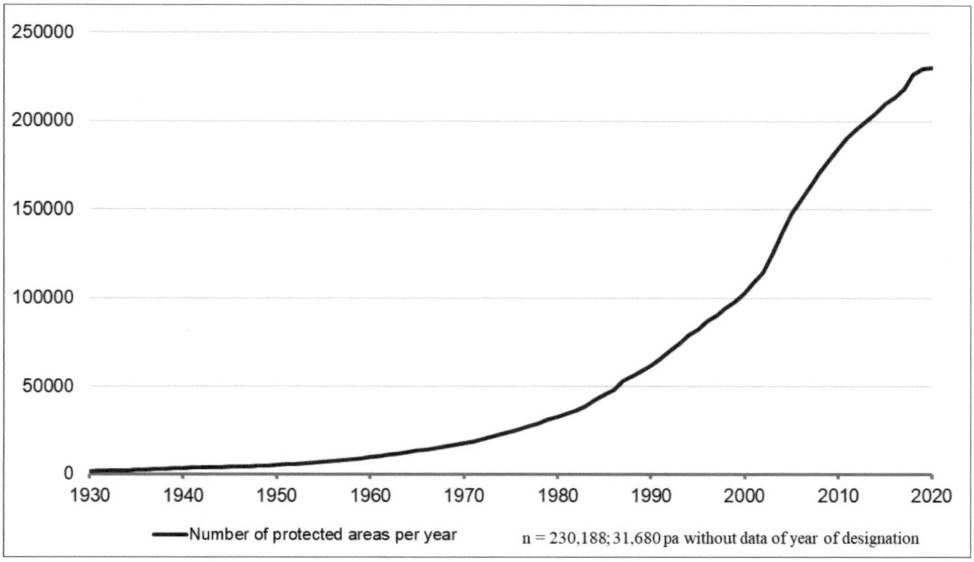

Source: Own illustration based on UNEP-WCMC et al., 2020

Over the past few decades, the number of protected areas has exponentially grown (see Figure 2). In 1962, the UN List of Protected Areas contained around 9,000 protected areas (DEGUIGNET et al., 2017, p. 2); by 2020, this number had increased to over 260,000 (UNEP-

WCMC et al., 2020). The collective actions of governments, publicly funded organizations and local communities led to rapid growth in the number of protected areas around the world (WATSON et al., 2014, p. 67). Most areas are on land; they protect over 20 million km², which is equivalent to around 15% of the earth's land surface. Although fewer in number, marine protected areas cover over 27.8 million km² and represent 7.7% of the world's oceans (see Figure 3).

Figure 3: Global distribution of protected areas in November 2020

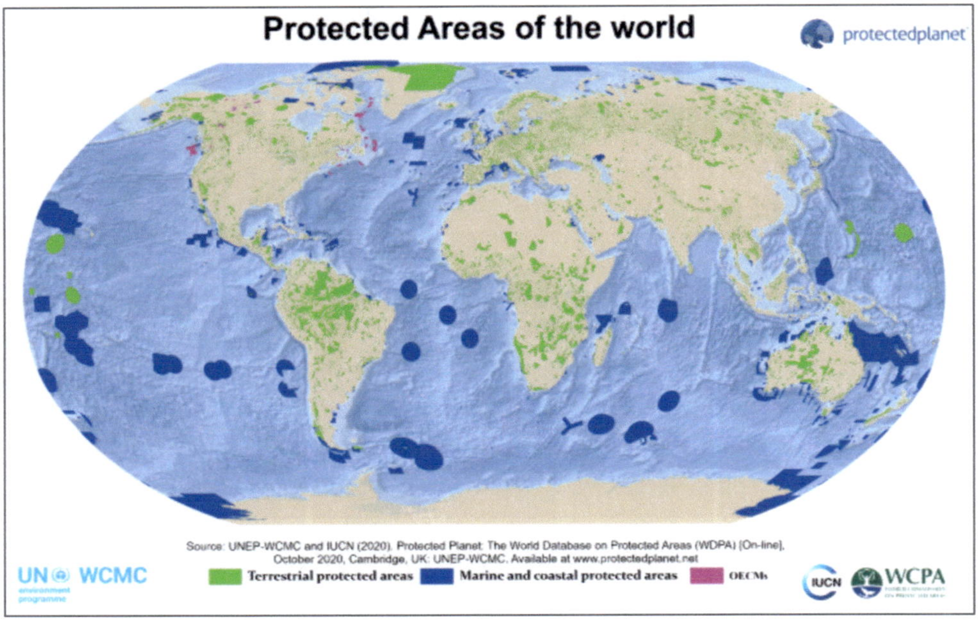

Source: UNEP-WCMC & IUCN, 2020

Protected areas are located all around world. However, the size of protected areas considerably varies between continents. Countries in Africa and South America and Australia, Greenland and Russia contain some very large protected areas. By contrast, Europe has a high number of small protected areas (see Table 2). The 160,684 protected areas in Europe cover a smaller area (3,719,236 km²) than the 8,448 protected areas in Africa (4,276,552 km²). In Latin America and the Caribbean, the average size of each protected area is by far the highest; 9,767 protected areas cover a remarkable 4,949,567 km² on land and 3,967,159 km² underwater (UNEP-WCMC, 2020a; 2020c; 2020d).

However, there are significant differences in the proportion of protected land between individual countries. Figure 4 shows the terrestrial coverage of protected areas across countries and territories. In Latin America, for example, Venezuela (54.1%) and Brazil (30.3%) have a considerably higher proportion of protected territory than Guyana (8.5%), Argentina (8.5%) and Uruguay (3.7%; UNEP-WCMC, 2020h;i;j;k;l). In Europe, European Union (EU) member states are notable for their high proportion of protected areas, which is undoubtedly linked to the Natura 2000 sites. The regions with the lowest percentage of protected areas are Central Asia, the Middle East and North Africa. With a few exceptions, the proportion is less than 5% in each country.

Table 2: Global comparison of total number and extent of protected areas as well as terrestrial and marine protected areas, in November 2020

	Protected areas	Protected land area in km²	Terrestrial protected area coverage	Protected marine area in km²	Marine protected area coverage
Africa	8,448	4,276,552	14.23%	1,736,583	11.63%
Asia and the Pacific	37,171	4,654,884	14.95%	11,176,454	18.31%
Europe	160,684	3,719,236	11.43%	1,378,626	7.86%
Latin America and the Caribbean	9,767	4,949,567	24.10%	3,967,159	17.32%
Polar	35	894,313	41.28%	3,964,822	44.78%
North America	44,542	2,218,301	11.41%	3,388,991	23.70%
West Asia	378	134,634	3.81%	15,865	1.10%

Source: Own illustration based on UNEP-WCMC, 2020a;b;c;d;e;f;g

Figure 4: Global overview of the percentage of protected areas across countries and territories

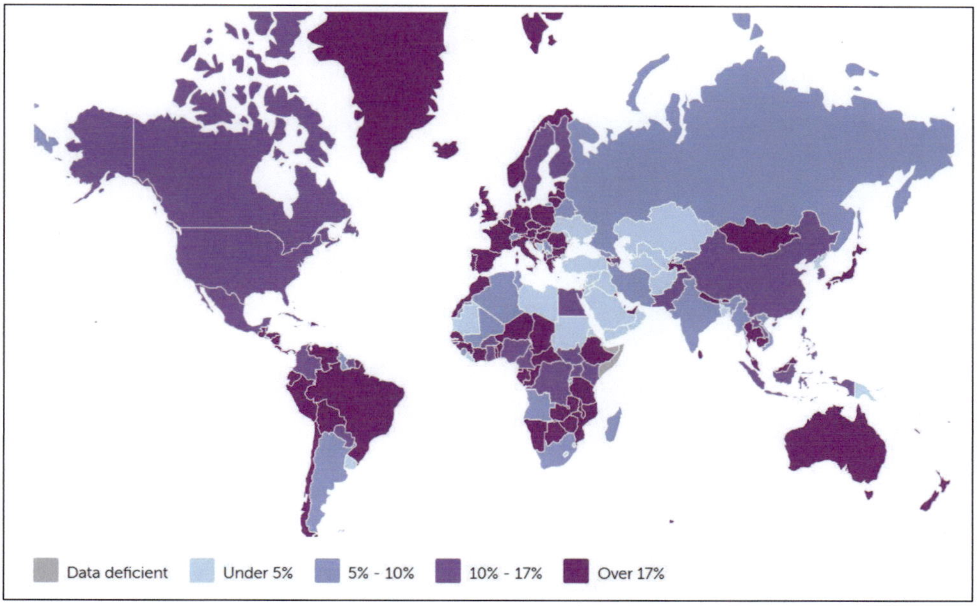

Source: UNEP-WCMC & IUCN, 2020

The total number and percentage of protected areas is continually changing as new areas are designated and existing areas are expanded. At the same time, however, some areas lose their designation as governments scale back or eliminate protections for previously protected zones. The observed increase in protected area coverage not only reflects new designations but also improved reporting by countries (UNEP-WCMC et al., 2018, p. 8). The legal extent of these designations varies across and within countries and regions and can be grouped into three broad categories: national designations created under national regulations, regional designations (e.g. Natura 2000 in Europe) and international designations, which are mostly created through conventions and agreements (e.g. World Heritage Sites, Ramsar sites or Biosphere Reserves). As a result, many protected areas are designated under more than one convention or legal instrument. DEGUIGNET et al. (2017) showed that almost a quarter of the world's protected areas are protected under more than one designation, with up to eight overlapping designations. These multiple designations occur in every region but are most common in Western Europe (pp. 2, 13-14).

3 The World Network of UNESCO Biosphere Reserves

3.1 Historic outline and idea of UNESCO Biosphere Reserves

The word 'biosphere' is derived from the German word *Biosphäre*, which is itself an amalgamation of the Greek words *bios* (mode of life) and *sphaira* (ball or sphere). The Austrian geologist Eduard Suess coined this term in his description of the earth's different spheres in 1875:

> 'One thing seems strange on this large celestial body formed of spheres, namely organic life. But even this is restricted to a certain zone, to the surface of the lithosphere. The plant which sinks its roots into the ground in search of food and at the same time lifts itself breathing into the air, is a good picture of the position of organic life in the region of interaction between the upper spheres and the lithosphere, and an independent biosphere can be distinguished on the surface of the solid'. (translated from German, SUESS, 1875, p. 159)

NEUMANN (2009) also described the biosphere as 'the totality of life on earth and the space in which it exists. One of the four components of the physical earth, including the lithosphere, hydrosphere, and atmosphere' (p. 308).

In nature conservation the term 'biosphere' gained importance in the late 1960s, which is also reflected in the 10-year International Biological Programme (1964–1974) on the co-ordination of large-scale ecological and environmental studies. ISHWARAN (2012) noted that the first moon landing with Apollo 11 in 1969 and the resulting ability to view Earth from outer space spurred ecologists to seek more holistic approaches to conservation and the use of natural resources (p. 93). In 1968, UNESCO convened the Intergovernmental Conference of Experts on the Scientific Basis for Rational Use and Conservation of the Resources of the Biosphere in Paris (commonly referred to as the Biosphere Conference), which was attended by 326 delegates from 63 nations. In the final report, the delegates articulated 20 recommendations; the first one called for an 'International Research Programme on Man and the Biosphere (…) of expanded and strengthened research, education and implementation on the problems of man and the biosphere', as 'many of the changes produced by man affect the biosphere as a whole (…) and that these problems cannot be solved on a regional, national or local basis, but require attention on a global scale' (UNESCO, 1968, pp. 2-3).

This was the kernel of UNESCO's MAB Programme, which was launched in 1971 and focused on

> 'the general study of the structure and functioning of the biosphere and its ecological divisions, on the systematic observation of, and research on, the changes brought about by man in the biosphere and its resources, on the overall effects of these changes upon the human species itself, and on the education and information to be provided on these matters'. (UNESCO, 1971, p. 7)

The MAB Programme was created as an interdisciplinary and international approach that centred on three themes: (1) the conservation of genetic resources and biological diversity, (2) international research and monitoring and (3) ecologically sustainable development (Coetzer et al., 2014, p. 83). As the responsible governance body, UNESCO established the MAB ICC, which decided to develop a network of environmentally significant sites that were representative of global ecosystems. In addition, they would support biodiversity conservation, meet the basic needs of local people and provide relevant ecological research. This proposal was incorporated into the concept of the Biosphere Reserves, which continue to be MAB's implementation instrument (Bridgewater, 2016, p. 2).

When the Biosphere Reserves were founded in 1974, UNESCO held a joint meeting with UNEP and IUCN to frame the programme's vision, mission and objectives. The three main objectives were (1) to conserve biological and genetic diversity and integrity, (2) to provide areas for ecological and environmental research and (3) to provide facilities for education and training. To achieve these objectives, a zonation system was introduced; it consisted of well-protected core area surrounded by one or several buffer zones along a gradient of decreasing use (UNESCO, 1974, p. 2; Batisse, 1982, p. 102; Bridgewater, 2016, p. 2). Initially, at the inception of the Biosphere Reserve network, the emphasis was on representative ecosystems rather than exceptional ones; the idea was to cover all of them as systemically and adequately as possible. The first global classification was based on Udvardy's (1975) concept of biogeographical provinces which were defined according to faunistic and floristic differences and vegetation structures. The classification identifies 193 biogeographical provinces belonging to 14 types of biomes within eight biogeographical realms. As a first approach, the objective was to establish at least one biosphere reserve in each biogeographical province (Batisse, 1982, pp. 104-105). From the beginning, the concept and vision of biosphere reserves were meant to be different from other international designations (Price et al., 2010, p. 550) and ahead of their time.

However, the reality of implementing the biosphere reserves did not always align with the concept. In the first two decades of MAB Programme, the concept of biosphere reserves significantly evolved. Three years after the first established biosphere reserve, there were already 118 biosphere reserves in 27 countries; by 1982, there were 214 biosphere reserves in 58 countries (Batisse, 1982, p. 105). However, these 'first-generation' biosphere reserves mainly prioritized the conservation function. They were superimposed on nationally designated protected areas; in addition, the links between environmental resources use and development and the usage of buffer zones for interactions with human settlements were mostly overlooked (Price et al., 2010, p. 550; Coetzer et al., 2014, pp. 84-85).

This mismatch between concept and reality was a key theme at the First International Biosphere Reserve Congress in 1983 in Minsk, as well as the potential to link conservation with human activities and sustainable development. The concept of a sustainable development and the term 'sustainable development' itself had recently been introduced by the World Conservation Strategy in 1980 (Price et al., 2010 p. 550). An output of the Minsk conference was the first Action Plan for Biosphere Reserves, which stated that people should be considered part of a biosphere reserve. In addition, the Scientific Advisory Panel on Biosphere Reserves recommended that biosphere reserves combine and harmonise three functions: conservation, development and logistic (research and monitoring). This would be achieved through three zones: the core area, a strictly delineated buffer zone and a transition area (Price, 2017, p. 31).

However, the first decade in the existence of biosphere reserves abruptly ended with the withdrawal of the United States of America, the United Kingdom and Singapore from UNESCO membership in 1985. Losing one quarter to one third of the MAB operational budget and the political backing of these countries brought immediate and continuing hardship (DYER & HOLLAND, 1988, p. 638; BRIDGEWATER, 2016, p. 3). Not only were the following years overshadowed by budget reductions, but many scientists also began to question their involvement in the programme and participants from the natural and social sciences focused on their own disciplines instead of engaging in interdisciplinary work. Another setback for the MAB Programme came with the re-organization of the World Heritage Convention within UNESCO. Previously, the MAB Secretariat had overseen both the natural World Heritage Sites and Biosphere Reserves; however, this advantage was lost when the World Heritage Centre was established in 1992 dealing with all World Heritage sites fully managed by the Culture Sector of UNESCO. Consequently, the MAB ICC decided that more attention should be given to the Biosphere Reserves programme and requested an Advisory Committee on Biosphere Reserves (BRIDGEWATER, 2016, p. 3). Another setback for the international visibility of the Biosphere Reserves programme was the IUCN's decision to no longer include biosphere reserves in their classification of protected areas. Whilst all biosphere reserves contain protected areas as core zones, their inclusion of human populations and settlements and their focus on sustainable development did not align with the IUCN's definition of a protected area (see chapter 2.4; PRICE, 2017, p. 31).

BRIDGEWATER (2016) described the 10-year period from 1985 to 1995 as a 'largely depressing decade of missed opportunities and reality checks' (p. 3). It ended, however, on a high note with the success of the second world conference of experts on biosphere reserves in Seville, Spain in 1995. The Seville Conference produced two major outputs: the Seville Strategy and the Statutory Framework for the World Network of Biosphere Reserves. The latter that defined the principles, criteria and designation procedures for biosphere reserves (BRIDGEWATER, 2016, p. 3). Both are milestones in the history of biosphere reserves and remain the most important guiding instruments for the MAB Programme. The documents outline actions to reconcile the conservation of biodiversity and biological resources through sustainable use. One important element was to define the structure of biosphere reserves as three spatial zones. Thus, the zonation that had already been addressed in 1974 and 1983 was clearly formalized. It consisted of a concentric system with a legally protected area at its core; this was dedicated to biodiversity conservation (see Figure 5). The core zone was surrounded by a buffer zone that allowed some level of human activity, as long as it was compatible with the site's ecological requirements (e.g. ecotourism) and provided the logistical function of research, education and training. Next, these two zones were surrounded by a transition area which included human settlements, and allowed cooperation between different actors and sustainable socio-economic development activities. Buffer and transition zones did not need to be legally protected (JOB et al., 2017, 1704-1706; UNESCO, 1996, p. 17). The Statutory Framework also included a periodic review process that would take place every 10 years, based on a report prepared by the concerned authority responsible for the biosphere reserve, to ensure every site's compliance with the biosphere reserve criteria (UNESCO, 1996, p. 18).

Figure 5: Zonation of UNESCO Biosphere Reserves

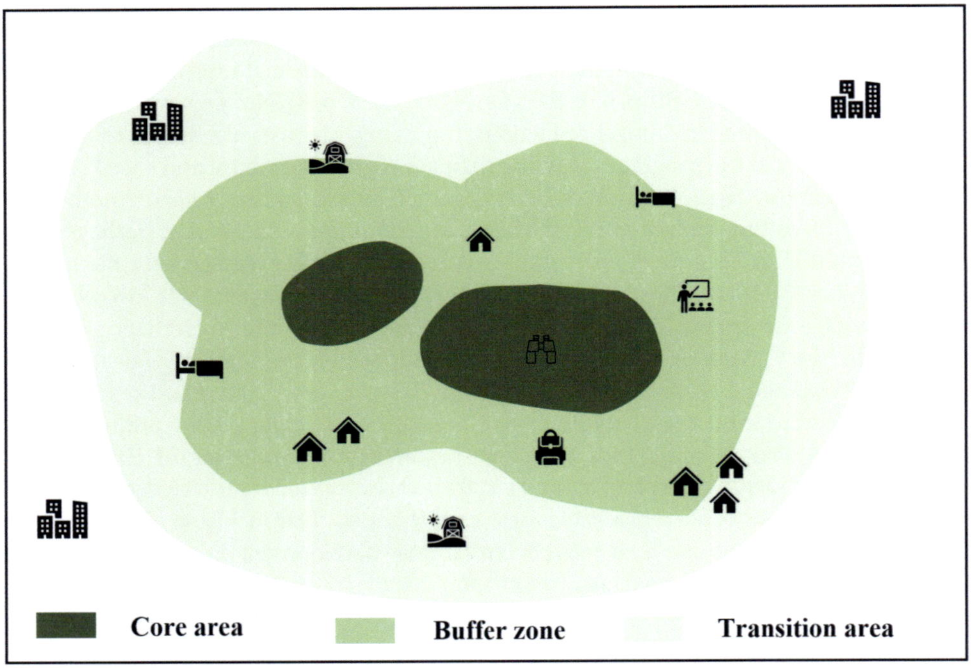

Source: Own illustration

In 2008, a third world conference on Biosphere Reserves was held in Madrid, which further strengthened the role of biosphere reserves as learning sites for sustainable development (BRIDGEWATER, 2016, p. 4). The resulting five-year Madrid Action Plan (MAP) noted that nearly all biosphere reserves designated after 1995 conform to the Statutory Framework for the World Network of Biosphere Reserves, but only 23% of sites designated from 1976 to 1984 included all three required zones. However, there was little new beyond the already widely accepted Statutory Framework, and the MAP did not establish any specific actions regarding the assessment of the process (PRICE et al., 2010, p. 551). BRIDGEWATER (2016) described the MAP rather as a reaffirmation and less as a renaissance of the MAB Programme (p. 4).

Over the last 10 years, the MAB Programme has focused on the quality of biosphere reserves rather than on quantity (JOB et al., 2017, pp. 1704-1706). The periodic review process included in the Statutory Framework in Seville to ensure that every biosphere reserve met the criteria moved increasingly into focus. Whilst governments submitted an increasing number of periodic reviews over the years, it became clear that a large proportion of biosphere reserves designated before 1996 did not include all three zones (COETZER et al., 2014, p. 85; PRICE, 2017, p. 31).

3.2 Periodic review process and the introduction of the Exit Strategy

The identified need to evaluate the implementation of the biosphere reserves resulted in the introduction of the periodic review process in 1995 as a part of the Statutory Framework. The MAB Secretariat described the periodic review process to take stock of progress made with the biosphere reserves, provide opportunities to update zonation and review objectives and management policies. The periodic review process was intended to allow the discussion of weak points and improvements in the quality of the biosphere reserves and their function as testing sites for sustainable development (UNESCO, 2017b). The Statutory Framework adopted by the MAB ICC and the General Conference of UNESCO states that

> 'the status of each biosphere reserve should be subject to a periodic review every ten years, based on a report prepared by the concerned authority, on the basis of the criteria of Article 4, and forwarded to the secretariat by the State concerned. The report will be considered by the Advisory Committee for Biosphere Reserves [MAB IAC] for recommendation to the ICC'. (UNESCO, 1996, p. 18)

Figure 6 shows each step of the periodic review process.

Figure 6: Seven steps of the periodic review process

The seven steps of the Periodic Review Process:

1. MAB Secretariat sends request for PR to Member State & National Committee
2. State sends PR to MAB Secretariat who transmits it to the IAC
3. IAC reviews the PR and makes recommendations to the MAB ICC
4. MAB ICC examines the PR as well as the recommendations and then decides

A **B**

not satisfactory status and sends recommendations for better compliance via the MAB Secretariat to the concerned state

5. After a 'reasonable period', an updated PR with evidence of corrective actions based on recommendations is expected
6. IAC reviews the updated PR and makes recommendations to ICC
7. MAB ICC makes final decision, either

satisfactory status unsatisfactory status

Source: Own illustration based on UNESCO, 1996

If the MAB ICC decides that the periodic review is unsatisfactory due to the biosphere reserve's substandard quality or lack of local capacity, the MAB IAC can recommend assistance from the responsible UNESCO Regional Office to prepare the periodic review (MATAR & ANTHONY, 2017, pp. 8-9). If the MAB ICC finds that the biosphere reserve still does not satisfy the criteria in Article 4 after a reasonable period, it can notify the director-general of UNESCO that it will no longer be part of the World Network of Biosphere Reserves. Alternatively, the concerned member state can recognize that a biosphere reserve under its jurisdiction does not satisfy the criteria of the Statutory Framework of UNESCO Biosphere Reserves, voluntarily remove it from the World Network of Biosphere Reserves and notify the MAB Secretariat (PRICE et al., 2010, p. 551).

During the third World Congress of Biosphere Reserves in Madrid in 2008, the periodic review process was widely discussed. The MAP stated that 98% of sites nominated after the Seville Strategy in 1995 adopted the three-zone scheme. However, only 23% of biosphere reserves designated between 1976 and 1984 and 65% of biosphere reserves designated between 1985 and 1995 had the three required zones (UNESCO, 2008, p. 9). The MAP also stated that the MAB Secretariat and the MAB Bureau should update the nomination and periodic review forms by 2010 and that all biosphere reserves should undertake periodic reviews and related actions to update zonation, management and other changes to meet the Seville and Madrid requirements and recommendations by 2013. In addition, the MAP stated that new biosphere reserves proposal should be supported through field-visits of regional and national experts to assure the implementation periodic review processes and compliance with the vision of post-Seville biosphere reserves (UNESCO, 2008, pp. 11, 15). However, PRICE et al. and BRIDGEWATER claimed that the MAP did not include strong and specific actions regarding the assessment of the periodic review process (PRICE et al., 2010, pp. 550-551; BRIDGEWATER, 2016, p. 4).

In a detailed evaluation of the periodic review process, PRICE et al. (2010) analysed all reports prepared by the MAB Secretariat and the final reports from MAB IAC meetings from 1996 to 2010. They found that one fifth of countries with biosphere reserves had never submitted a periodic review report. Most of these biosphere reserves were designated before 1986, and the most important factor of missing periodic review reports was whether countries were sufficiently committed to the World Network of Biosphere Reserves and had effective national institutions and processes in place. Notably, by 2010, the United States of America, had not submitted a single periodic review report for its 47 biosphere reserves, all of which were designated before 1995. In addition, Bulgaria had not submitted any periodic review reports for its 16 biosphere reserves by 2010 (PRICE et al., 2010, p. 552).

In 2011, an expert group composed of members of the MAB IAC, representatives of the Austrian MAB National Committee and the MAB Secretariat recommended to the MAB ICC that 'those pre-Seville designated biosphere reserves having a core zone for research and protection only and which cannot meet the criteria of the Seville Strategy by 2013 should be withdrawn from the World Network of Biosphere Reserves' (UNESCO, 2011a, p. 26).

During the 24th session of the MAB ICC in 2012, several member states underlined the importance of the period review process and the need to specify deadlines to meet the Statutory Framework criteria. The MAB ICC requested the MAB Secretariat to send a letter to all concerned member states with copy to UNESCO Permanent Delegations, asking them to inform the Secretariat on their plan for undertaking the periodic review process and updating biosphere reserves to meet the Statutory Framework criteria by the next ICC session. The letter should include a reference to the Statutory Framework, specify the possibility for member states to withdraw a site and remind them of the Secretariat's availability to support them with the submission of the periodic review (UNESCO, 2012a, p. 38).

In 2013, at the 25th session of the MAB ICC, the MAB Secretariat presented an update on the periodic reviews. By then, 287 biosphere reserves had completed a periodic review. Only six biosphere reserves fully met the criteria, 80 sites satisfactorily met the criteria, 138 partly met the criteria, 55 did not meet the criteria and six biosphere reserves were recommended withdrawal from the network. The MAB Secretariat also stated that 112 biosphere reserves had never done a periodic review and that nine countries had never completed a periodic review for any of their biosphere reserves. Therefore, as requested by the MAB ICC the previous year, the MAB Secretariat sent 102 letters to member states to request periodic reviews and follow up on recommendations issued by the MAB ICC. In total, 145 biosphere reserves replied in time about how and when they would submit a periodic review report; however, 131 biosphere reserves did not answer the letters sent out in October 2012 by the MAB Secretariat (UNESCO, 2013, pp. 37-38). After numerous delegates commented on this development and possible next steps, the MAB ICC decided to adopt an Exit Strategy:

> 'A biosphere reserve of the MAB Programme is an attractive designation that not only serves to enhance conservation but also sustainable development and research throughout the world. As a consequence, the number of biosphere reserves has increased considerably from 391 sites in 94 countries in the year 2000 to 621 biosphere reserves in 117 countries (including 12 transboundary sites) in 2013. The MAB Programme as a scientific programme has also evolved since its inception in 1971, and so have methods, competencies, experience, and knowledge developed on how to apply the biosphere reserve concept in practice. In this context the MAB Programme has started a process to ensure the continued adherence of the sites established as biosphere reserves to the objectives of their establishment and to ensure the credibility and coherence of the World Network of Biosphere Reserves (...)'. (UNESCO, 2013, pp. 38-39)

Consequently, the MAB ICC established a three-step process to manage the periodic review process and assess, monitor and improve the quality of the World Network of Biosphere Reserves. First, the MAB ICC sent warning letters to the MAB National Committees and the concerned biosphere reserves and copied UNESCO National Commissions, ministries and the concerned countries' permanent delegations, providing a three-month period for responses. In case there was no reply, the Exit Strategy recommended that the MAB ICC question and decide whether to remove the site from the World Network of Biosphere Reserves. If a reply was received from the MAB National Committee, the MAB Secretariat placed the concerned biosphere reserve on an internal pending list and provided a new deadline of one year for the submission of a periodic review report (UNESCO, 2013, p. 39).

In direct response to the warning letters, the number of received periodic reviews considerably increased (MATAR & ANTHONY, 2017, p. 12). On the periodical reviews, most MAB IAC recommendations focused on the links between conservation and sustainable development, concerns about small spatial extent of a biosphere reserve to fulfil all functions and incomplete or inefficient zonation, insufficient community involvement and inadequate management plans for the biosphere reserve. At the beginning of the Exit Strategy at least 22 biosphere reserves did not have a permanent population and others did not have the spatial zonation underlined by the Statutory Framework (COETZER et al., 2014, pp. 85-86). The greatest challenges identified in the preparation of the period reviews were a lack of technical and financial capacity. A lack of human resources needed to implement required recommendations at the site level and to report and monitor progress was a limiting factor for compliance. Moreover, the cost of preparing the periodic review and possible expert fees could be high in some countries (MATAR & ANTHONY, 2017, p. 12). PRICE et al. (2010) tried to quantify related costs and concluded that, although they were not comparable across countries, they varied from USD 2,200 in Canada, where time was donated by national experts, to USD 43,000 in France (p. 552).

When reflecting on the evaluation of management effectiveness and the implementation of the biosphere reserve concept, it is important to note that there was also a discussion about self-evaluation bias. Mainly due to a lack of financial resources, it was not possible to conduct external reviews or missions to evaluate the biosphere reserves, as in other programmes or international designations. In an analysis of periodic reviews, MATAR and ANTHONY (2017) stated that bias can result in either defensive or counter-defensive attributions, such as the inflation of success by biosphere reserve managers if they feel that the period review is directly linked to their job performance or the understatement of their success to attract more resources for management. In addition, the accuracy of expert opinions can greatly vary depending on their understanding of the biosphere reserve concept. This factor is highly reliant on the evaluators selected for the periodic review, especially if the latter is only conducted by internal members of the biosphere reserve structure and hired consultants (p. 17).

During the Exit Strategy process, the fourth World Congress of Biosphere Reserves was held in Lima in March 2016. It resulted in the Lima Action Plan for 2015–2025, which recommended that the World Network of Biosphere Reserves develop and strengthen models for sustainable development, communicate experiences and lessons learned, support evaluation and high-quality management strategies, develop polices for sustainable development and planning and help member states and stakeholders to urgently meet the SDGs by sharing experiences (UNESCO, 2016a, p. 2). PRICE (2017) formulated four requirements for achieving these high goals. First, stakeholders must be made aware that particular areas are biosphere reserves and have potential to help them meet individual, local, regional and global priorities. Second, the involvement of diverse stakeholders in activities is key to ensure participation from the preparation of a targeted policy or plan to its implementation. Third, each biosphere reserve requires a participatory governance structure that involves a wide range of stakeholders and has support from all levels of governance. PRICE argued that this support was most evident in countries where biosphere reserves were recognized in national legislation, such as Australia, Brazil, Estonia, France, Germany, Kyrgyzstan, Spain and Ukraine (p. 38). ISHWARAN et al. (2008) concluded that this is a critical challenge, as 80% of designated biosphere reserves do not fall under any protected areas legislation. For the most part, biosphere reserve managers have no jurisdiction beyond the core zone (i.e.

the buffer or transition zones). It became clear that the identification of authorities with the mandate and resources to coordinate stakeholder interests in biosphere reserves would be key to the success of the World Network of Biosphere Reserves (pp. 123-125). In the MAP, Goal 11.1 also encouraged member states to include biosphere reserves in their own legislation by 2013 (UNESCO, 2008, p. 16). The fourth requirement, which was intimately linked to the third, was the adequate provision and allocation of financial resources to biosphere reserves to meet the various goals (PRICE, 2017, p. 38).

In addition, the 10-year reporting timeline for the periodic reviews has long been discussed. Some argue that this period is too long to effectively monitor changes in biosphere reserves and to tackle the responding recommendations of the MAB ICC and MAB IAC. Whilst five years might be too short to implement MAB ICC recommendations such as changes in zonation, a 10-year period seems too long for adaptive management (PRICE et al., 2010, p. 555). After serious discussions within the IAC, the idea of reducing the timeframe from 10 to five years was mostly abandoned due to limited capacity to review twice the number of periodic reviews (MATAR & ANTHONY, 2018, pp. 13-14).

Since the inception of the MAB Programme, there have also been discussions about the use of the word 'reserve'. In many languages, the term has a negative connotation, as it can convey an area in which people are excluded (STOLL-KLEEMANN & O'RIORDAN, 2018, p. 350). In some federal states in Germany, biosphere reserves are also called 'biosphere areas' under state law; in Austria, they are called 'biosphere parks' (MUKE BADEN-WÜRTTEMBERG, 2020; ÖSTERREICHISCHE UNESCO-KOMMISSION, 2020). In Japan, the Japanese National Committee for MAB calls biosphere reserves 'UNESCO Eco Parks' (MATSUDA et al., 2015, p. 244; KOHSAKA & MATSUOKA, 2015, p. 4). Therefore, STOLL-KLEEMANN and O'RIORDAN (2018) argued that the term 'biosphere reserves' should be replaced with 'biosphere landscapes' or 'biosphere regions', as this would better reflect the positive aspects of these institutions (p. 352).

During the MAB ICC meeting in 2016 in Lima, the MAB Secretariat stated that the final deadline to submit of all pending periodic review reports and follow-up information on recommendations was the end of September 2015. However, the MAB ICC's final decision on the Exit Strategy was postponed until its next meeting in Paris in 2017; therefore, all countries and sites that still did not meet the criteria would have until 30 September 2016 to reply. During the sessions, delegates shared challenges and concerns about preparing the periodic reviews, such as the time needed for participative review processes and the involvement of local stakeholders and language issues (UNESCO, 2016b, p. 99).

One year later, the response rate from the 270 sites in 75 countries affected by the Exit Strategy was very high, and all countries sent a response. Many biosphere reserves had improved zonation, governance and management within the given timeframe (UNESCO, 2017a, p. 15). However, the chair of the MAB Bureau stated that 95 biosphere reserves were still affected by the Exit Strategy. A long discussion was held during the MAB ICC meeting, mainly on the political consequences and implications of a possible MAB ICC decision to remove sites and on the possible loss of reputation to the MAB Programme if a clear decision was not taken. During the session, many delegates also asked the MAB ICC to consider providing additional time to countries and biosphere reserves, to taking into account sociocultural differences and the need for a follow-up process to further improve the quality of the World Network of Biosphere Reserves (UNESCO, 2017a, pp. 16-17). Ultimately, the MAB ICC decided to allow more time to complete the Exit Strategy, setting a deadline for 2020, and to institute the Exit Strategy with a

> 'Process of excellence and enhancement of the World Network of Biosphere Reserves as well as quality improvement of all members of the World Network, to ensure that they serve as models for the implementation of the 2030 Agenda and its SDGs'. (UNESCO, 2017a, p. 17)

The Process of Excellence and Enhancement was supported by an ad hoc working group to further develop it, prepare input for discussion and to allow a decision to be taken about the sites still pending approval by the 32nd session of the MAB ICC in 2020 (UNESCO, 2018, p. 33). The working group comprised two MAB ICC members for each UNESCO Regional Group, the chairperson of the MAB ICC and the chairperson of the IACBR. During the 30th session of the MAB ICC in 2018, the MAB Secretariat stated that 64 biosphere reserves in 31 countries were still affected by the Process of Excellence and Enhancement (UNESCO, 2018, p. 34) and that most sites did not meet the criteria due to zonation and governance issues (UNESCO, 2018, p. 32). The ad hoc working group's preliminary results were presented at the 31st session of the MAB ICC in 2019. The group recommended that MAB National Committees be established and to conduct voluntary and informal midterm reviews every five years. This would enable performance monitoring by national-level authorities between periodic reviews. The ad hoc working group also highlighted the importance of regional networks in supporting biosphere reserve reviews and facilitating peer-to-peer exchange and support. After some discussion, the MAB ICC asked the working group to continue its work and further develop the idea of a short and easy review mechanism coordinated by the MAB National Committees or national MAB focal point every five years after the site's designation or last periodic review report (UNESCO, 2019, pp. 19-21).

The 32nd session of the MAB ICC was postponed from June 2020 to a virtual session on 27 and 28 October 2020 due to the COVID-19 pandemic. As a result of challenges and delays associated with the pandemic, the MAB Bureau and MAB Secretariat decided to provide more time to the countries and biosphere reserves affected by the Exit Strategy and the ad hoc working group's proposal development. In 2020, 45 biosphere reserves were still part of the Process of Excellence and Enhancement (UNESCO, 2020a, pp. 1-2).

The process formerly known as the Exit Strategy and now known as the Process of Excellence and Enhancement was completed at the MAB ICC meeting in Abuja, Nigeria in September 2021 (UNESCO, 2022).

So far, the implementation of the Exit Strategy is seen by many scholars and MAB ICC delegates as a great success because it ensures quality control and strengthens the World Network of Biosphere Reserves (Köck & Arnberger, 2017, p. 88; Price, 2017, p. 38; UNESCO, 2018, pp. 32-33). However, many local communities did not want to be involved in the rezonation process; in some sites, there were no local communities in or around the biosphere reserve to include in a transition zone with an emphasis on sustainable development (Price, 2017, pp. 31-33). Since the inception of the MAB Programme in the 1970s, 13 UNESCO member states have voluntarily withdrawn 59 biosphere reserves from the World Network of Biosphere Reserves. 51 biosphere reserves (86%) were withdrawn after the adaptation of the MAP in 2008 and 46 sites after the Exit Strategy was implemented in 2013.

The most withdrawals were made by the United States of America (19), followed by Australia (12), Bulgaria and the United Kingdom (eight each), Austria (four), and the Central African Republic, the Democratic Republic of Congo, Denmark, Germany, Mexico, the Netherlands, Norway and Sweden (one each; see Table 3).

Table 3: List of biosphere reserves that were voluntarily withdrawn from the World Network of Biosphere Reserves

Country	Biosphere reserve	Year of designation	Year of withdrawal
Australia	Southwest	1977	2002
	Macquarie Island	1977	2011
	Prince Regent	1977	2018
	Yathong	1977	2018
	Hattah-Kulkyne and Murray-Kulkyne	1981	2018
	Wilson's Promontory	1981	2018
	Barkindji	2005	2018
	Croajingolong	1977	2020
	Kosciuszko	1977	2020
	Unnamed	1977	2020
	Uluru	1977	2020
	Riverland	2004	2020
Austria	Gossenköllersee	1977	2014
	Gurgler Kamm	1977	2014
	Unter Lobau	1977	2016
	Neusiedler See	1977	2016
Bulgaria	Maritchini Ezera	1977	2002
	Doupkata	1977	2017
	Kamtchia	1977	2017
	Koupena	1977	2017
	Alibotouch	1977	2020
	Doupki-Djindjirtza	1977	2020
	Mantaritza	1977	2020
	Parangalitza	1977	2020

Country	Biosphere reserve	Year of designation	Year of withdrawal
Central African Republic	Bamingui Bangoran	1979	2019
Democratic Republic of Congo	Lufira	1982	2020
Denmark	North-east Greenland	1977	2019
Germany	Bayerischer Wald	1981	2007
Mexico	Islas del Golfo de California	1995	2020
The Netherlands	Waddensea Area	1986	2018
Norway	Northeast Svalbard	1976	1997
Sweden	Lake Torne Area	1986	2010
United Kingdom	Caerlaverock	1976	2002
	Isle of Rhum	1976	2002
	St. Kilda	1976	2002
	Claish Moss	1977	2002
	Taynish	1977	2010
	Moor House Upper Teesdale	1976	2012
	Loch Druidibeg	1976	2013
	North Norfolk Coast	1976	2014
United States of America	Aleutian Islands	1976	2017
	Beaver Creek	1976	2017
	California Coast Ranges	1983	2017
	Carolinian South Atlantic	1986	2017
	Central Plains	1976	2017
	Coram	1976	2017
	Desert	1976	2017
	Fraser	1976	2017
	H.J. Andrews	1976	2017
	Hubbard Brook	1976	2017
	Konza Prairie	1978	2017
	Land Between the Lakes	1991	2017

Country	Biosphere reserve	Year of designation	Year of withdrawal
United States of America	Niwot Ridge	1979	2017
	Noatak	1976	2017
	Stanislas-Tuolumne	1976	2017
	Three Sisters	1976	2017
	Virgin Islands	1976	2017
	San Dimas	1976	2018
	San Joaquin	1976	2019

Source: UNESCO, 2020c; UNESCO, 2020a; UNESCO, 2020b

3.3 Spatial diffusion of the World Network of UNESCO Biosphere Reserves

This section provides an overview of the spatial-temporal development and diffusion of the World Network of Biosphere Reserves from its first designation in 1976 to 2020. Scientists have different views on how to divide over 40 years of UNESCO Biosphere Reserves into groups. Many researchers (COETZER et al., 2014, p. 85; REED, 2016; p. 450; STOLL-KLEEMANN et al., 2010, p. 227) described the Seville Conference and its key outputs, the Seville Strategy and the Statutory Framework, as a major paradigm shift in the MAB Programme, with a clear division between pre- and post-Seville biosphere reserves. Therefore, they refer to pre-Seville sites as first-generation biosphere reserves and post-Seville sites as second-generation biosphere reserves. ISHWARAN et al. (2008) subdivided the pre-Seville phase into two generations: from 1976 to the first action plan for biosphere reserves in 1984 and the following years up to the second world conference in Seville in 1995 (p. 123). Meanwhile, JOB et al. (2017) distinguished five phases according to the international conferences on biosphere reserves which took place in 1983, 1995, 2008 and 2016: the establishment phase (1976-1982), the pre-Seville phase (1983-1994), the post-Seville phase (1995-2007), the rapid expansion or 'boom' phase (2008-2014) and the delivery phase (2015 to the present; pp. 1704-1706). However, given the timeframe of a post-Seville biosphere reserve designation, which follows a bottom-up approach and includes extensive stakeholder involvement, a new designation typically takes several years; therefore, it is challenging to define global phases of less than 10 to 15 years. Thus, in this dissertation, the division of biosphere reserves follows the idea of two generations: a pre-Seville generation with sites that mainly focused on a science-led approach to nature conservation in core areas and a post-Seville generation with sites that mainly focus on sustainable development, feature buffer and transition zones and actively include human activities.

The overall objective of the Exit Strategy is to transform all first-generation biosphere reserves into post-Seville biosphere reserves and models for sustainable development through rezonation and revised management; after the completion of the Exit Strategy, there should be a single consolidated World Network of Biosphere Reserves, with all sites conforming to the Statutory Framework.

As of the 32nd session of the MAB ICC virtual meeting in October 2020, the World Network of Biosphere Reserves includes 714 biosphere reserves in 129 countries, including 21 transboundary biosphere reserves (UNESCO, 2020c).

The biosphere reserves designated in each generation are shown in a map, and the most important characteristics of each phase are compiled in some overview tables (see Figures 7-9; Tables 4-10). The spatial data for each site were downloaded from the WDPA of the UNEP World Conservation Monitoring Centre (UNEP-WCMC) and visualized using ESRI ArcMap software. However, around 150 UNESCO Biosphere Reserves were not included in the WDPA; most of them were designated after 2012. The locations of the missing biosphere reserves were manually included in the geographic information system (GIS) database by searching for their coordinates in the UNESCO database of the World Network of Biosphere Reserves, which is accessible on the UNESCO MAB website (UNESCO, 2020c). The locations were double-checked on OpenStreetMap. The complete list of biosphere reserves and their respective year of designation was also derived from the UNESCO database of the World Network of Biosphere Reserves (UNESCO, 2020c).

First-generation biosphere reserves: conservation focus (1976–1995)

REED (2016) described the nomination and selection process for first-generation biosphere reserves based on representative ecosystems, which followed biologist Miklos UDVARDY's (1975) international classification of biogeographical provinces. This approach focused on the representativeness of biosphere reserves rather than the uniqueness of biological features to understand and address widespread environmental challenges across a diversity of landscapes (p. 450). The selection process was often referred to as the selection of test sites for research projects (KÖCK & ARNEGGER, 2017, p. 86). Most biosphere reserve designations were superimposed on existing protected areas or research sites, which resulted in top-down and half-hearted governance structures. As a result, the biosphere reserve label was largely an add-on, without any effects on the wider landscape or management (PRICE et al., 2010, p. 550; MATAR & ANTHONY, 2018, pp. 9-10; BRIDGEWATER, 2016, p. 3; WINKLER, 2019, p. 173). In fact, more than 80% of all biosphere reserves affected by the Exit Strategy were first-generation biosphere reserves (UNESCO, 2017a, p. 15).

By June 1995, 324 biosphere reserves had been designated in 82 countries (UNESCO, 1996, p. 3). Although this number suggests an equal global distribution of first-generation biosphere reserves, it is important to note that 54.9% were located in Europe and North America, followed by 16.7% in Asia and the Pacific (see Table 5). The United States of America hosted 47 biosphere reserves (14.5% of the total number), followed by the Russian Federation with 22 sites (see Table 4). Among the 10 countries with the most biosphere reserves, seven were located in Europe or North America. ISHWARAN et al. (2008) stated that the 1970s vision of biosphere reserves as sites for conservation and interdisciplinary research appealed more to industrialized countries than developing economies at that time

(p. 123). Moreover, the political situation of the Cold War, the struggle for power in the new international environment and nature conservation programmes, conventions and organizations played a significant role in the diffusion of first-generation biosphere reserves.

Table 4: Top 10 countries with the most first-generation biosphere reserves

Country	Number
United States of America	47
Russian Federation	22
Bulgaria	17
Spain	13
Australia	11
Germany	11
Mexico	11
China	10
Iran	9
United Kingdom	9

Source: UNESCO, 2020c

Table 5: Distribution of first-generation biosphere reserves by UN regions

UN region	Number	Percentage
Africa	34	10.5%
Arab States	10	3.1%
Asia and the Pacific	54	16.7%
Europe and North America	178	54.9%
Latin America and the Caribbean	48	14.8%

Source: UNESCO, 2020c

The world map in Figure 7, which shows the distribution of first-generation biosphere reserves, illustrates well the global distribution of biosphere reserves. However, it also shows the concentration of sites in Europe and North America. By contrast, the lowest representation of first-generation biosphere reserves was in the Arab States, South Asia and Southern Africa.

Figure 7: Spatial allocation of first-generation Biosphere Reserves, designated between 1976-1995

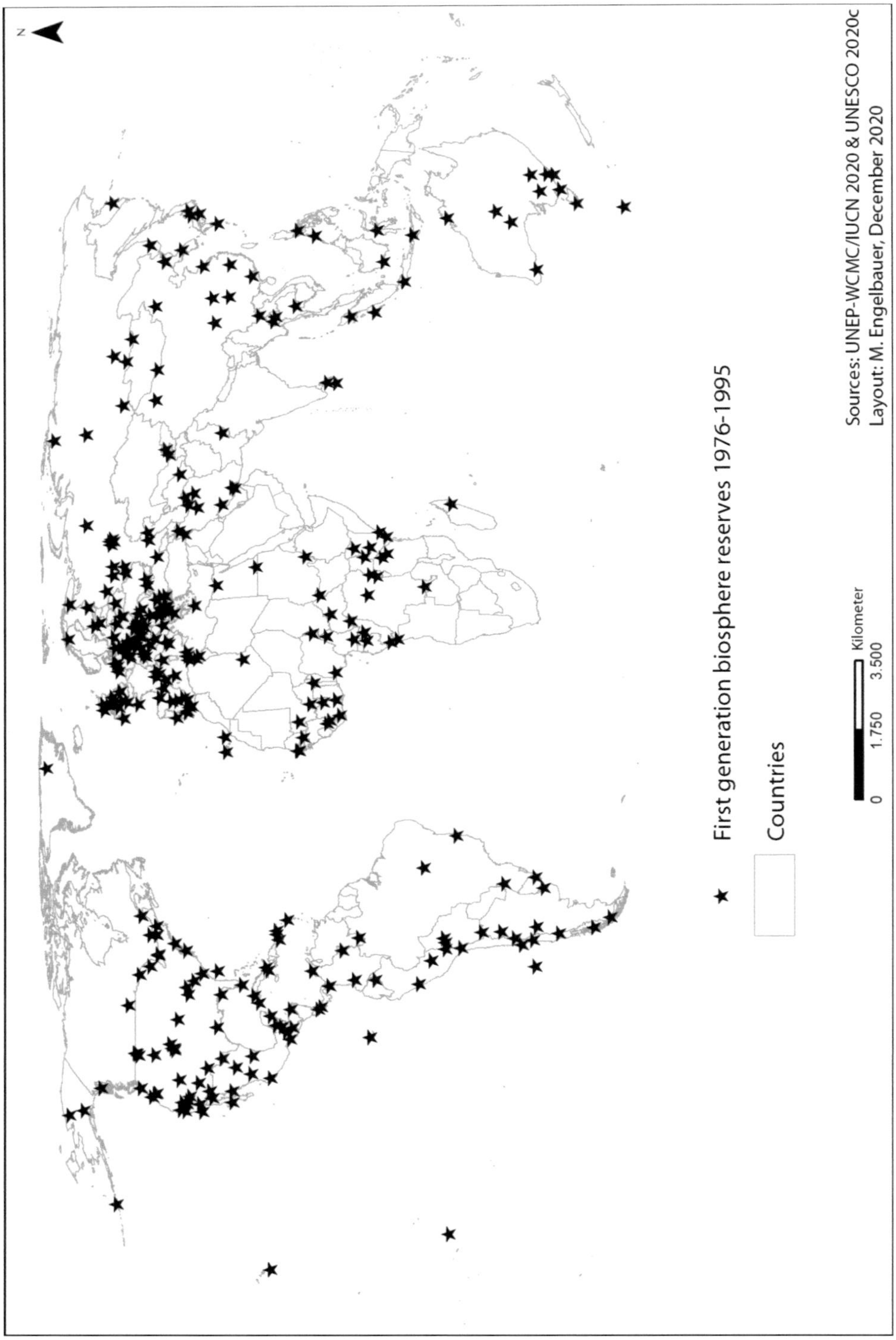

Second-generation biosphere reserves: sustainable development focus (1996–2019)

After the Seville conference, second-generation biosphere reserves reflected a shift in focus from a research-driven to a management-driven programme and from protection of individual areas to ecosystem or regional conservation with sustainable development as an overarching priority (COETZER et al., 2014, pp. 84-85; REED, 2016, p. 451). The aim of this integrated approach was to create 'more than just protected areas' (UNESCO, 1996, p. 5). Development was at the core of second-generation biosphere reserves, with the goal of fostering linkages between humans and ecosystems. In addition, clear changes to the structure of the biosphere reserves were made over time: (1) the designation of several core areas rather than only one, (2) the boundary delineation of transition zone(s) and (3) the greater integration of all functions in all zones, which means that conservation, sustainable development and logistical support can be implemented in all zones but to varying degrees (MATAR & ANTHONY, 2017, p. 4). Whilst the implementation of the new vision continued to be challenging for local communities and decision makers in zonation, land tenure, monitoring, governance and coordination mechanisms for moderating stakeholder interests, the link between conservation and development appealed to many policymakers and decision makers. It seemed especially attractive to countries in the developing world (ISHWARAN et al., 2008, p. 124).

Between 1996 and 2020, 454 new UNESCO Biosphere Reserves were designated in 111 countries (see Figure 8). However, the most new sites were in Europe and North America (38.1%), but the decrease of 16.8% in comparison to the first generation is contributing mainly Asia and the Pacific, which comprised 26.2% of new second-generation biosphere reserves, followed by Latin America and the Caribbean (18.5%). Moreover, the Arab States and Africa gained around two percentage points each compared to the number of the first-generation biosphere reserves. However, the Arab States still had the lowest percentage of total global designations at 5.1% (see Table 7). Given ISHWARAN et al.'s (2008) hypothesis that the designation of second-generation biosphere reserves was particularly attractive to developing countries, it is somewhat surprising that the country with the most new designations was Spain, with 38 sites. In the top 10 countries with the most new designations, Spain was followed by Mexico (31); Russian Federation (29); China (24); Italy (16); Indonesia (13); Canada, India, and Kazakhstan (12 each); and Portugal. Among the 20 countries with the most new designations, half were located in the Global South. According to the 2019 Human Development Index (HDI), 12 were countries with very high human development, seven were countries with high human development and one was a country with medium human development (see Table 6).

Figure 8 shows the global diffusion of second-generation biosphere reserves. The concentration of new sites in Mediterranean countries and in Mexico is clear. Notably, there were no new designations in the United States of America. Compared to the map of first-generation biosphere reserves, the new designations were in regions that had been underrepresented in the list of first-generation biosphere reserves: Eastern Africa, Southern Africa and South Asia. In general, it can be stated that second-generation biosphere reserves were distributed in many countries around the world.

Table 6: Top 20 countries of total number of new designations of biosphere reserves in the second generation

Country	Number	North-South divide	HDI ranking in 2019
Spain	38	North	Very high: 0.904
Mexico	31	South	High: 0.779
Russian Federation	29	North	Very high: 0.824
China	24	South	High: 0.761
Italy	16	North	Very high: 0.892
Indonesia	13	South	High: 0.718
Canada	12	North	Very high: 0.929
India	12	South	Medium: 0.645
Kazakhstan	12	South	Very high: 0.825
Portugal	11	North	Very high: 0.864
South Africa	10	South	High: 0.709
Argentina	9	South	Very high: 0.845
Vietnam	9	South	High: 0.704
France	7	North	Very high: 0.901
Republic of Korea	7	North	Very high: 0.916
Sweden	7	North	Very high: 0.945
Mongolia	6	South	High: 0.737
Japan	6	North	Very high: 0.919
Germany	6	North	Very high: 0.947
Algeria	6	South	High: 0.748

Source: Own illustration based on UNDP, 2020, pp. 343-345; UNESCO, 2020c

Table 7: Distribution of second-generation biosphere reserves by UN regions

UN region	Number	Percentage
Africa	55	12.1%
Arab States	23	5.1%
Asia and the Pacific	119	26.2%
Europe and North America	173	38.1%
Latin America and the Caribbean	84	18.5%

Source: UNESCO, 2020c

Figure 8: Spatial allocation of second-generation Biosphere Reserves, designated between 1996-2020

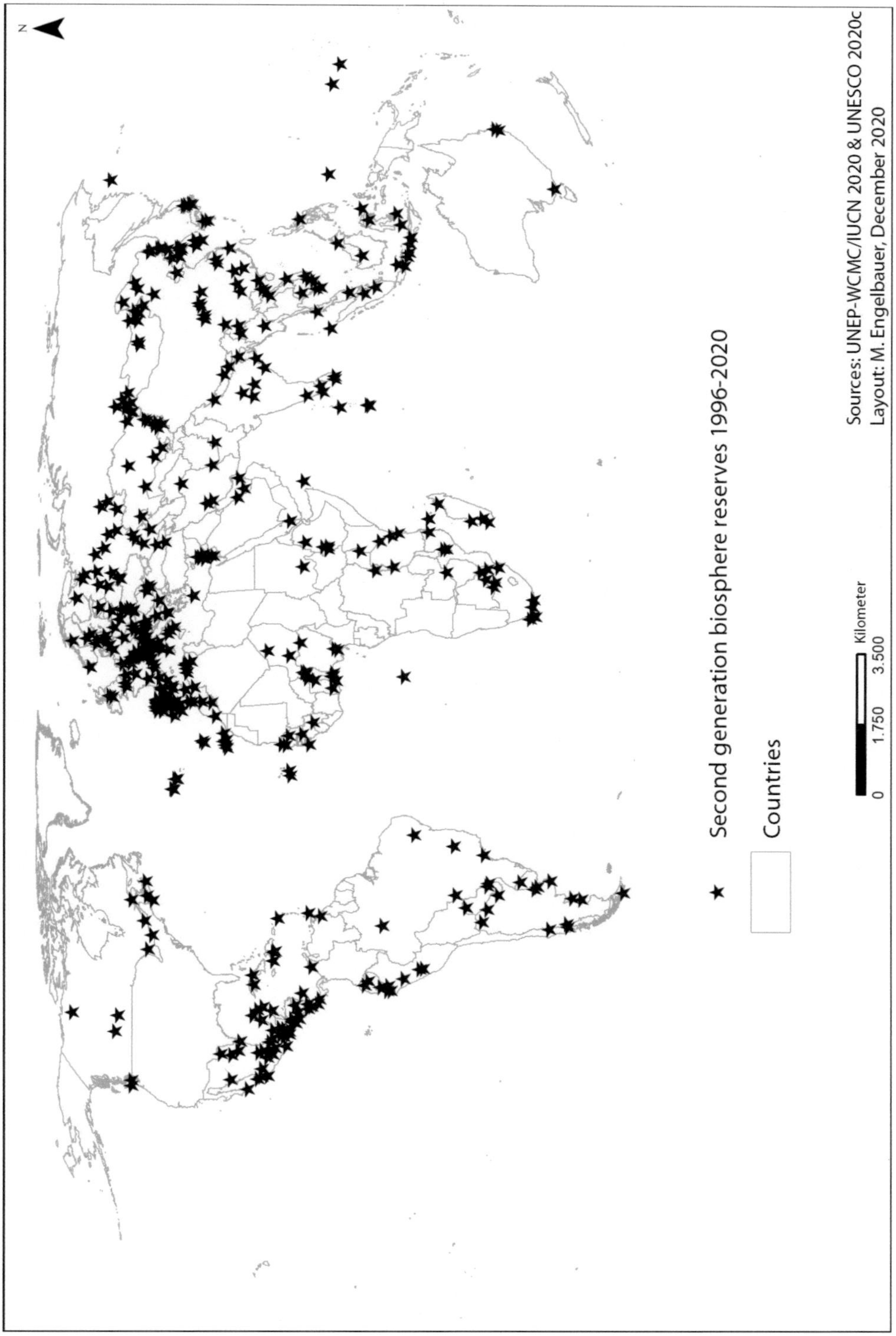

Consolidated World Network of Biosphere Reserves (2020 onwards)
Since October 2020, the World Network of Biosphere Reserves includes 714 biosphere reserves in 129 countries, including 21 transboundary sites (UNESCO, 2020c).

More than 50 years after the establishment of the UNESCO Biosphere Reserves programme, modernized first-generation biosphere reserves and second-generation biosphere reserves have become a truly global network (see Figure 9). Some spatial clusters can be identified in Mexico and Central America, in the Mediterranean (especially Spain, Portugal and Italy) and in the Balkan states. In Asia, there is a higher concentration of sites around the Sea of Japan in the Korean states, Japan, eastern Russia and China. However, New Zealand, Papua New Guinea and many Pacific Island states are not yet represented in the World Network of Biosphere Reserves, and the number of Australian biosphere reserves has significantly decreased over the past few decades. In Africa, spatial representation has significantly improved in recent years, but 18 countries on the African continent still do not have any biosphere reserves (BOTHA et al., 2021). In particular, none of the four neighbouring countries in southwest Africa have any biosphere reserves: Angola, Botswana, Namibia and Zambia.

In 2020, five countries accounted for 202 biosphere reserves, which represents over a quarter of sites in the entire network. Spain led with 52 sites, followed by the Russian Federation (47), Mexico (41), China (38) and the United States of America (28; see Table 8; Brenner & Job, 2021). The 20 countries with the most biosphere reserves accounted for 373 sites, which represents more than half of the total number of biosphere reserves in the World Network. Except for the Arab States, all UN regions were represented. However, Western and developed countries still dominate the World Network of Biosphere Reserves; 42.5% of biosphere reserves are in Europe or North America, and 81.1% are in countries with very high or high human development (see Table 9 and Table 10).

A closer analysis of the biosphere reserves' total surface area reveals significant differences between countries. Whilst Spain had the most biosphere reserves, they were comparatively small, with an average size of 143,375 ha; the total area of all Spanish biosphere reserves was only 74,555 km^2 (see Table 8). By comparison, biosphere reserves in the Russian Federation occupied the largest total area worldwide at 500,366 km^2, which corresponded to an average biosphere reserve size of 1,064,609 ha. By total surface area, the next countries in the rankings were Indonesia (293,197 km^2), United States of America (266,793 km^2), Canada (227,305 km^2) and Chile (118,123 km^2). On average, Indonesia had the largest individual biosphere reserves at 1,543,144 ha, followed by Canada (1,262,806 ha), Chile (1,181,229 ha), South Africa (1,176,719 ha) and the Russian Federation (1,064,609 ha). By comparison, the average size of European biosphere reserves was small, but the host countries' total surface areas were also correspondingly smaller. In terms of the total surface area of biosphere reserves in relation to total national surface area, 23.18% of Portugal's national territory was designated as biosphere reserves. Chile follows in the 20 countries with the highest share of biosphere reserves surface area to national surface area in 2020 (15.62%), followed by Indonesia (15.34%), Spain (14.74%), Vietnam (13.24%) and South Africa (9.64%). Amongst large countries, the total surface area of biosphere reserves occupied a significantly lower share of total national surface area: 2.93% for the Russian Federation, 2.71% for the USA, 2.28% for Canada, 1.76% for India and 0.66% for China.

Table 8: Top 20 countries with the most biosphere reserves in 2020

Country	Number of biosphere reserves	Total surface area of biosphere reserves [km2]	Average surface area of biosphere reserves [ha]	Total surface area of biosphere reserves to total national surface area [%]
Spain	52	74,555	143,375	14.74%
Russian Federation	47	500,366	1,064,609	2.93%
Mexico	41	118,015	287,840	6.01%
China	34	63,818	187,699	0.66%
United States of America	28	266,793	952,832	2.71%
Indonesia	19	293,197	1,543,144	15.34%
Italy	19	26,691	140,478	8.84%
Canada	18	227,305	1,262,806	2.28%
Germany	16	20,858	130,361	5.84%
Argentina	15	109,745	731,636	3.95%
France	14	44,285	316,319	8.03%
Iran	13	57,414	441,649	3.53%
India	12	57,732	481,101	1.76%
Kazakhstan	12	67,214	560,118	2.47%
Portugal	12	21,378	178,148	23.18%
Poland	11	13,114	119,219	4.19%
Chile	10	118,123	1,181,229	15.62%
Japan	10	13,469	134,687	3.56%
South Africa	10	117,672	1,176,719	9.64%
Vietnam	9	43,805	486,723	13.24%

Source: Own illustration based on UNESCO, 2020c; UN, 2020b

Figure 9: Spatial allocation of UNESCO Biosphere Reserves in 2020

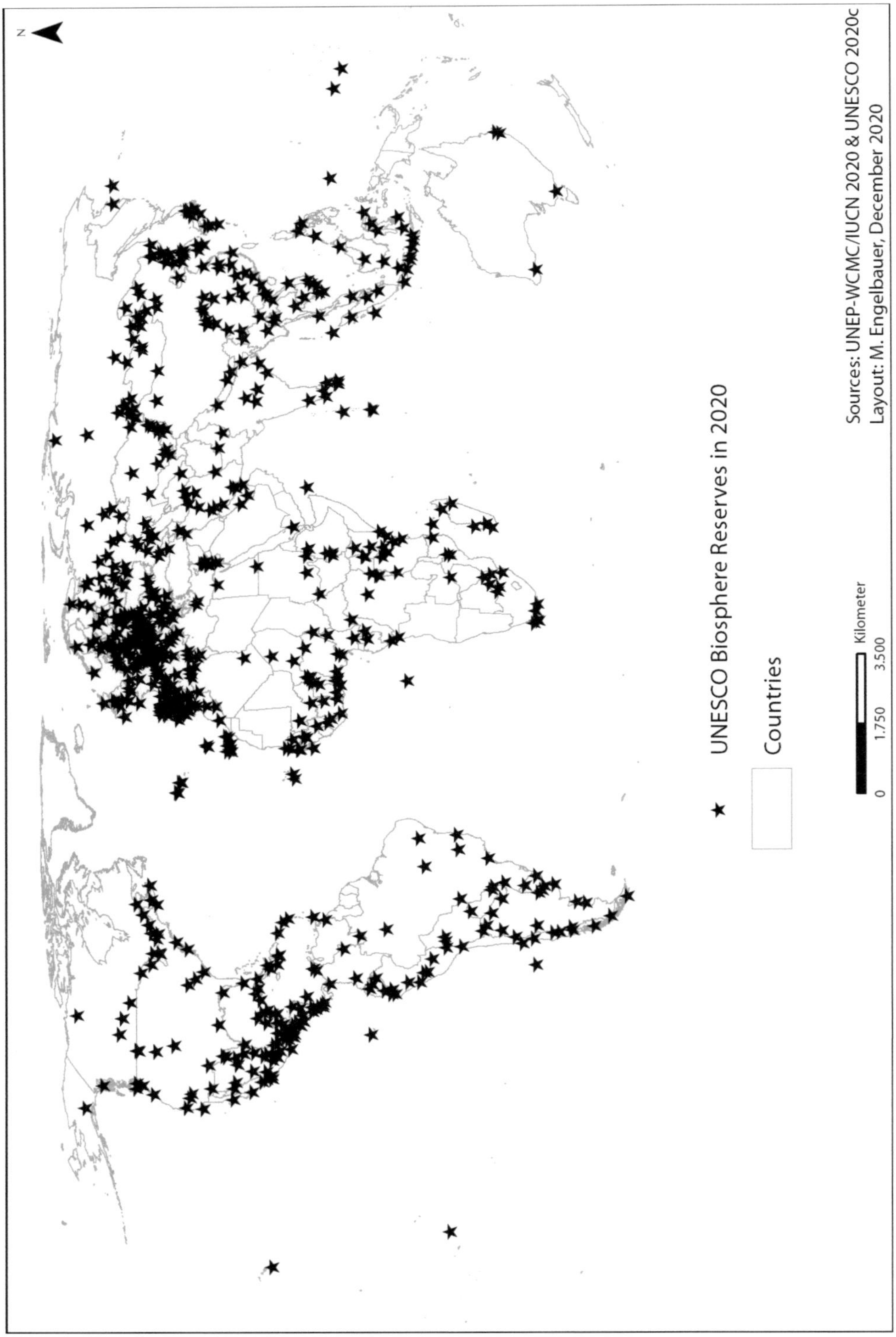

Table 9: Biosphere reserves in 2020 sorted to Human Development Index classes

Human Development Index	Number of biosphere reserves	Percentage
very high human development	357	50.0 %
high human development	222	31.1 %
medium human development	67	9.4 %
low human development	63	8.8 %
no data	5	0.7 %

Source: Own illustration based on UNDP, 2020, pp. 343-345

Table 10: Biosphere reserves in 2020 divided into UN regions

UN region	Number	Percentage
Africa	85	11.9 %
Arab States	33	4.6 %
Asia and the Pacific	161	22.6 %
Europe and North America	303	42.5 %
Latin America and the Caribbean	131	18.4 %

Source: UNESCO, 2020c

3.4 Key actors and structure of the World Network of Biosphere Reserves

Given the MAB Programme's intergovernmental nature, UNESCO provides it with a framework to support national governments in the planning and implementation of research and training programmes as well as technical and scientific advice. The MAB Secretariat plays a key role in the implementation of the strategy and roadmap, which are adopted by the member states, and in the management of the World Network of Biosphere Reserves. It acts as an intermediary between the governing bodies of the UNESCO MAB Programme, the member states and the biosphere reserves themselves. The MAB Secretariat is based at UNESCO's Division of Ecological and Earth Sciences in Paris, and its secretary is also currently the director of the latter. For all programme activities, the MAB Secretariat works closely with different field offices around the world to coordinate the MAB Programme at the national and regional levels. The MAB Programme is funded through UNESCO's regular budget and mobilized funds-in-trust granted by UNESCO member states, bilateral and multilateral sources, and extra-budgetary funds provided by countries, the private sector and private institutions (UNESCO, 2019, p. 2; PRICE et al., 2010, p. 554).

The main governing body of the MAB Programme is the MAB ICC. It consists of 34 UNESCO member states, which are elected at UNESCO's biennial General Conference. Normally, half of MAC ICC members end their terms of office, and new council members are elected at the biennial General Conference. The outgoing members are usually replaced by members from the same regional group, but they can also be re-elected. The MAB ICC meets every year, usually at the UNESCO headquarters in Paris or elsewhere at the invitation of a MAB ICC member state (e.g. 2018 in Palembang, Indonesia; 2016 in Lima, Peru; 2014 in Jönköping, Sweden; and 2011 in Dresden, Germany). During the MAB ICC meetings, every member state only has one vote, but they can send as many experts or advisers as they wish. Other UNESCO member states that are not members of the MAB ICC can send representatives as observers. International organizations such as the IUCN and UN agencies such as UNEP, Food and Agriculture Organization, UNDP, World Meteorological Organization and World Health Organization are also invited. The principal tasks of the MAB ICC are to

> 'guide and supervise the MAB Programme, review the progress made in the implementation of the Programme (Secretariat report and reports of MAB National Committees), recommend research projects to countries and to make proposals on the organization of regional or international cooperation, assess priorities among projects and MAB activities in general, co-ordinate the international cooperation of Member States participating in the MAB Programme, co-ordinate activities with international scientific programmes, and consult with international non-governmental organizations on scientific or technical questions'. (UNESCO, 2020d)

The MAB ICC also decides on new biosphere reserves in the World Network of Biosphere Reserves and takes note of recommendations on periodic review reports. During its meetings, the MAB ICC elects a chairperson and five vice-chairpersons for the MAB Bureau. Its members are nominated from each of the UNESCO's geopolitical regions; between MAB ICC meetings, the authority of the council is delegated to the MAB Bureau. From 2018 to 2020 (and thus during the field work undertaken for this dissertation), the MAB Bureau consisted of Indonesia as chair; Estonia, Honduras, Nigeria and Sudan as vice-chairs; and Sweden as rapporteur. Since 2020, Nigeria held the chair of the MAB Bureau; the Russian Federation, Peru, Kazakhstan and Sudan have been vice-chairs; Austria has been the rapporteur (UNESCO, 2020d). Figure 10 illustrates the structure of the MAB Programme and its key actors.

A fourth important body in the MAB Programme is the MAB IAC which is the primary scientific and technical committee within the World Network of Biosphere Reserves. It advises the MAB ICC and the director-general of UNESCO in all matters concerning nominations, changes and withdrawals related to biosphere reserves, periodic reviews and the development, operation and monitoring of the network. Twelve members are appointed for four years by the director-general and selected in consultation with member states for their scientific qualifications and experience in promoting and implementing biosphere reserves (UNESCO, 2011b, pp. 1-2). The MAB IAC plays a significant role in the discussions about the Exit Strategy, the Process of Excellence and Enhancement and the periodic review, as it examines all periodic reviews and presents recommendations to the MAB ICC.

In addition, a second advisory body, the International Support Group to the MAB Programme (MAB ISG), provides recommendations to the MAB Secretariat. Its purpose is to advise the MAB Secretariat on the implementation of relevant MAB action plans and strategies. Participation is open to all member states with delegations at UNESCO headquarters. The MAB ISG meets once or twice per year to receive first-hand information from the MAB Secretariat for member states and advises the MAB Secretariat on ongoing challenges in the implementation of biosphere reserves in various countries (UNESCO, 2020d).

Figure 10: Structure of the MAB Programme and key actors

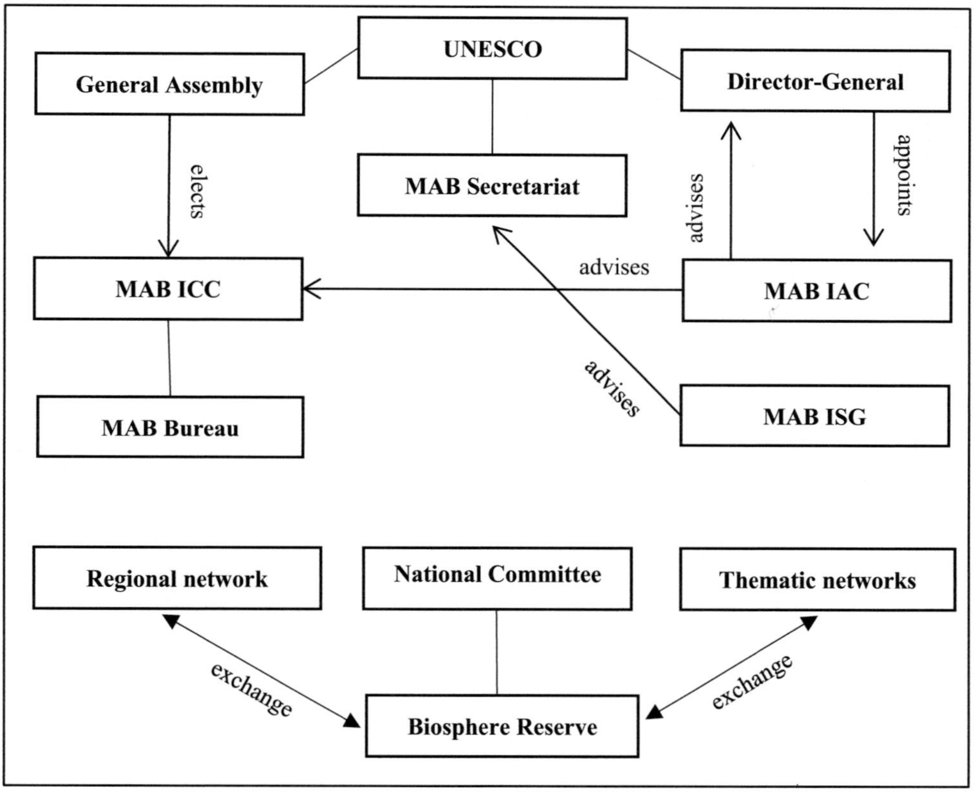

Source: Own illustration

A key advantage of a global network is the opportunity to learn and benefit from others' experiences. In the World Network of Biosphere Reserves, sub-networks have emerged over several decades based on commonalities and shared interests to exchange lessons learned from the implementation of biosphere reserves. These networks are also an alternative to UNESCO's hierarchical and diplomatic governance system and enable direct exchanges between biosphere reserve managers and stakeholders without intermediaries in Paris or national capitals.

Sub-networks in the World Network of Biosphere Reserves can be divided into two broad groups: regional and the ecosystem-specific networks. The regional sub-networks include AfriMAB for Africa; ArabMAB for the Arab States; EuroMAB for Europe and North America; Ibero-American MAB for Latin America and the Caribbean, Portugal and Spain; EABRN for East Asia; Mediterranean BR Network for the Mediterranean; NordMAB for Nordic countries; PacMAB for the Pacific; SACAM for South and Central Asia; SeaBRnet for Southeast Asia; and REDBIOS for the East Atlantic or Macaronesia region (UNESCO, 2020d; UNESCO, 2019, p. 1; UNESCO, 2017, pp. 12-13). They all aim to foster the exchange of knowledge and experiences and promote regional collaboration between biosphere reserves. Meanwhile, ecosystem-specific (or thematic) sub-networks provide valuable insights on research, capacity building and educational collaborations between all biosphere reserves engaged in research on a specific ecosystem. To date, there are seven thematic sub-networks for drylands; mangroves; marine, coastal and island areas; mountains; savannahs; tropical forests; and wetlands (UNESCO, 2020d).

A major challenge related to the international UNESCO designation is the fact that all biosphere reserves remain under the jurisdiction of member states. It is the member state's responsibility to designate sites to or withdraw them from the World Network of Biosphere Reserves and to ensure their appropriate governance and management by the relevant governing institutions, which can be public institutions, private institutions, NGOs or a combination of these (MATAR & ANTHONY, 2018, p. 8). To promote biosphere reserves and national participation in the MAB Programme and the MAB ICC meetings, UNESCO member states should establish a permanent and fully functioning national committee. Ideally, these government-appointed MAB National Committees should liaise with UNESCO National Commissions, the MAB focal points for the MAB Secretariat and different national institutions and ministries involved in the MAB Programme (UNESCO, 1997, pp. 1-2). The MAB Programme currently operates through 158 MAB National Committees established in the 195 member states and nine associate member states of UNESCO. In some countries, the establishment of MAB National Committees is still in process; in such cases, they are represented by their UNESCO National Commission (UNESCO, 2020d; UNESCO, 2019).

3.5 Interlinkages with other UNESCO programmes for nature conservation

As mentioned in Section 2.5, there have been several parallel attempts to undertake nature conservation and protect key habitats. Even within UNESCO, there are three different approaches and designations for the conservation and protection of nature: the popular natural UNESCO World Heritage Sites, UNESCO Biosphere Reserves and, since 2015, UNESCO Global Geoparks (see Figure 11). This dissertation only examines the interlinkages between the three UNESCO designations, as they have similar advantages and disadvantages related to their common administrative structures, use of the brand in tourism and the visibility and risk of confusion with regard to the use of the UNESCO name and logo.

The Convention Concerning the Protection of the World Cultural and Natural Heritage, commonly known as the World Heritage Convention, is an international agreement adopted at the General Conference of UNESCO in 1972. It is based on the premise that 'parts of the cultural and natural heritage are of outstanding interest and therefore need to be preserved as part of the world heritage of mankind as a whole' (UNESCO, 1972 p. 1). Natural World Heritage Sites are considered according to the following criteria:

> 'natural features consisting of physical and biological formations or groups of such formations, which are of outstanding universal value from the aesthetic or scientific point of view; geological and physiographical formations and precisely delineated areas which constitute the habitat of threatened species of animals and plants of outstanding universal value from the point of view of science or conservation; natural sites or precisely delineated natural areas of outstanding universal value from the point of view of science, conservation or natural beauty.' (UNESCO, 1972, p. 2)

In addition to the cultural and natural World Heritage Sites, it is also possible to nominate mixed cultural and natural World Heritage Sites if they satisfy in part or in whole the definitions of both cultural and natural heritage in the convention (UNESCO WHC, 2019, p. 20). In recent decades, natural World Heritage Sites have not only developed into symbols of conservation but also revenue generator through international tourism (HALL & PIGGIN, P. 2001; BUCKLEY, 2004; BORGES et al., 2011; ENGELS et al., 2011; SU & LIN, 2014). They are also increasingly seen as engines for sustainable development (CONRADIN et al., 2014; CONRADIN et al., 2015).

To implement the World Heritage Convention, the General Assembly elects the World Heritage Committee which consists of 21 members and meets at least once a year. The Committee's main tasks are to make final decisions on the inscription and removal of properties on the World Heritage List, to examine of the state of conservation of the World Heritage Sites through processes of reactive monitoring and periodic reporting and to define the use of the World Heritage Fund and the allocation of financial assistance based on requests from states parties (UNESCO WHC, 2019, pp. 12-13). The Committee is assisted by a Secretariat appointed by the director-general of UNESCO which is assumed by the World Heritage Centre and three advisory bodies, two for the cultural heritage sites and IUCN advising for the natural World Heritage Sites (UNESCO WHC, 2019, pp. 14-15).

In 1997, it was agreed that states parties would provide periodic reports on the application of the convention and the state of conservation of World Heritage Sites. The periodic reporting is a self-reporting process and must be led and submitted by the state's parties. The reports are prepared on a regional basis according to a pre-established six-year cycle. Therefore, state parties from one of the five UN regions complete and submit a periodic report every year (UNESCO WHC, 2020a; UNESCO WHC, 2019, pp. 55-56).

As of June 2020, the World Heritage List included 1,121 World Heritage Sites, of which 869 were cultural, 213 were natural and 39 were mixed (UNESCO WHC, 2020b).

The concept of geoparks was developed in Europe in the late 1980s. The idea was to combine a particular geological heritage and a sustainable territorial development strategy. The term 'geopark' was first used in Germany in relation to the Gerolstein District Geopark, which was established in 1989; it was enlarged and renamed Vulkaneifel Geopark in 2000 (HENRIQUES & BRILHA, 2017, pp. 349-351). An important starting point for the placement of geological heritage on the international agenda for nature conservation was the International Declaration of the Rights of the Memories of the Earth in Digne, France in 1991. Since 1992, the director of UNESCO's Division of Earth Sciences had favoured the idea of geoparks (EDER, 1999; XUN & MILLY, 2002, p. 33). With his support, UNESCO decided to promote a global network of geosites at the 29th session of the UNESCO General Conference in 1997. Two years later, the UNESCO Global Geoparks programme was proposed to support national initiatives to preserve important geological sites in line with the principles of sustainable development (PATZAK & EDER, 1998, p. 33). However, in 2001, the executive board of UNESCO proposed to the director-general not to pursue the development of a UNESCO geosite or geoparks programme but rather to support efforts of member states as appropriate (HENRIQUES & BRILHA, 2017, p. 350).

Under the auspices of UNESCO, 17 existing European geoparks in the newly created European Geoparks Network and eight selected Chinese national geoparks were combined to form the Global Network of National Geoparks (GGN) in 2004. This worldwide network of 25 geoparks encouraged other countries, including Iran and Brazil, to develop national geoparks programmes (JONES, 2008, p. 273). In 2011, seven years after the establishment of the GGN, a new attempt was made to formalize a UNESCO geoparks programme. The General Conference of UNESCO called on the director-general to 'examine the feasibility of establishing a possible UNESCO geoparks programme or initiative, building on the existing success and experience of the Global Geoparks Network and geoparks' (UNESCO, 2012b, p. 34).

Figure 11: Overview and key facts about UNESCO labels for nature conservation

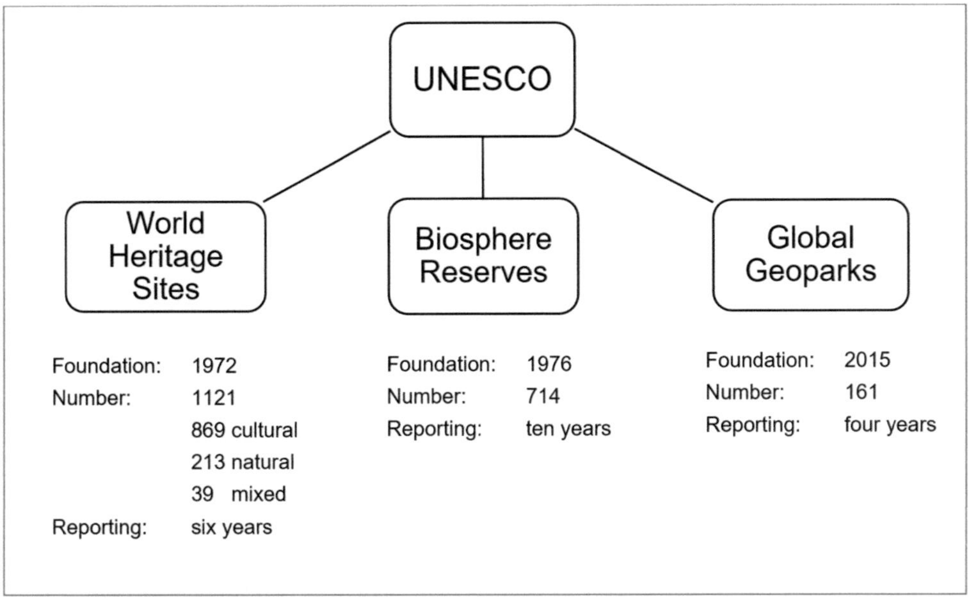

Source: Own illustration based on UNESCO, 2017b; UNESCO, 2020c; UNESCO WHC, 2019; UNESCO WHC, 2020b; UNESCO, 2017c; UNESCO, 2020c

In 2015, UNESCO finally approved the International Geoscience and Geoparks Programme, an umbrella initiative that comprised the current International Geoscience Programme and the UNESCO Global Geoparks. The new label aimed to establish a mechanism of international cooperation by which geological heritage areas with international value could support each other to engage with local communities and promote awareness of geological heritage and adopt a sustainable approach to the area's development (HENRIQUES & BRILHA, 2017, pp. 352-353).

As with the World Heritage Sites and UNESCO Biosphere Reserves, UNESCO Global Geoparks are subject to regular quality control. Every four years, they must undergo a thorough revalidation process to examine their functioning and quality. First, a self-evaluation and progress evaluation must be submitted by each geopark. Then, a field mission is undertaken by two evaluators. Depending on the results of the field evaluation report, the geopark may continue to operate for another four-year period (UNESCO, 2017c).

As of July 2020, there are 161 UNESCO Global Geoparks in 44 countries. The countries with the most UNESCO Global Geoparks were China (41 sites), Spain (15 sites), Italy (nine sites), Japan (nine sites), France and UK (seven sites each; UNESCO, 2020e).

As described in Section 2.4, many protected areas are designated under more than one convention or legal instrument (DEGUIGNET et al., 2017). There are not only multiple protected area designations with national and international designations but also awards of several international labels for nature conservation in the same place. Even the UNESCO designations sometimes overlap with each other and create MIDAs (SCHAAF & CLAMOTE RODRIQUES, 2016). PRICE et al. showed that World Heritage Sites and UNESCO Biosphere Reserves overlap in 65 sites in 42 countries (PRICE et al., 2010, p. 550), JOB et al. identified 79 natural World Heritage Sites that are also embedded in the World Network of Biosphere Reserves (JOB et al., 2017, p. 1712) and SCHAAF and CLAMOTE RODRIQUES determined that 109 Biosphere Reserves overlap with 100 World Heritage Sites, 16 Biosphere Reserves overlap with 14 Global Geoparks and 15 Global Geoparks overlap with 13 World Heritage Sites. There were even two sites with all three UNESCO designations (pp. 4-5).

Biosphere Reserves and World Heritage Sites have different objectives; World Heritage Sites must be of outstanding universal value, whilst Biosphere Reserves must represent distinctive natural and cultural landscapes and focus on regional sustainable development. However, according to JOB et al., these protected area categories can be complementary. The integration of the two concepts can be promising, as they are both managed by UNESCO; in addition, natural World Heritage Site can feature the core zones of biosphere reserves. Furthermore, the buffer and development zones may enable the sustainable management of the wider surrounding area of the World Heritage Site. Prominent examples can be found in Mesoamerica: the Mariposa Monarca Biosphere Reserve and World Heritage Site in Mexico and the Maya Biosphere Reserve and the Tikal World Heritage Site in Guatemala (JOB et al., 2017, pp. 1712-1713).

MATAR and ANTHONY (2017) argued that the multitude of designations emphasizes the importance of the sites, but their management and evaluation is more complex due to multiple layers of governance and institutional requirements (p. 3). PRICE et al. (2010) also indicated that the multiple designations are reflected in additional reporting processes; despite calls, they still lack a comparative review of these processes to increase harmonization and improve information sharing (p. 550). Whilst continued membership in Global Geoparks is subject to a periodic review every four years, World Heritage Sites require reporting every six years and Biosphere Reserves have the longest interval, with a 10-year periodic review cycle. Regarding the content of the reports, the basic information required is similar, but other details can substantially differ from one UNESCO designation to another. For example, Global Geoparks require detailed evidence on how sites and their managers have contributed to the work of the network, whilst the periodic review form for Biosphere Reserves requires detailed and analytical information on the fulfilment on the site's conservation and development functions, governance status and management system. Therefore, the three secretariats must collaborate on reporting duties and share assessments with each other and their governing bodies to avoid situations in which a site has fulfilled all requirements under one designation but failed to do so for another (SCHAAF/CLAMOTE RODRIGUES 2016, pp. 53-54).

4 Methodology

4.1 Qualitative expert interviews

Expert knowledge and expertise play a central role in modern societies. This is limited not only to the areas of science and technology but also practically all areas of life, such as where to study and questions of nutrition, health and child rearing. This knowledge dependence in many areas of life and work has contributed to the knowledge society that we live in today (BOGNER et al., 2014, p. 10).

Regarding the use of expert interviews as a scientific method for gathering information, it is important to acknowledge the methodological debate about where the constituted expert obtains their recognition as an expert (BOGNER et al., 2014, p. 11). Some researchers argue that the expert is the construct of our research interest as we give them the attribution of an expert by our specific research interest and our request for the interview (MEUSER & NAGEL, 2009, p. 18; PFADENHAUER, 2009, pp. 89-90). However, BOGNER et al. (2014) indicated that the expert alone cannot be considered a construct, as we refer to social conventions in the process of selecting the expert. Experts are mostly people in prominent social positions; for example, the are identifiable as experts because they sit on an advisory board or have the title of professor. Therefore, an expert is defined by a specific research interest and the social standing of the expert; in other words, they are a construct of the researcher and society. 'Experts can be understood as persons who - based on specific practical or experiential knowledge relating to a clearly limited problem area - have created the possibility of structuring the concrete field of action in a meaningful way and in a way that guides the actions of others' (BOGNER et al., 2014, p. 13, translated from German). Thus, expert knowledge is particularly effective in practice and provides orientation and guidance to others (p. 14).

Expert interviews can shorten time-consuming data gathering processes, particularly those related to practical insider knowledge. This is especially relevant in cases in which it is difficult or impossible to gain access to a particular social field. If the targeted expert is willing to participate in the research and holds a key position at an organization, there is the opportunity for expanding the researcher's access to the field. With support from an expert in a key position, the researcher may find it easier to gain access to an extended circle of experts. Furthermore, the expert may suggest additional potential interviewees with expertise in the research topic during the interview itself. In addition, expert interviews offer the researcher an effective and rapid means of obtaining relevant results. Similar scientific backgrounds can increase the level of motivation of the expert to participate in an interview. Secondary motivating factors may include awareness of the scientific and/or political relevance of their field or personal achievements, the desire to help make a difference, professional curiosity about the topic and an interest in sharing thoughts and ideas with an external person (BOGNER et al., 2009, p. 2).

In this research project, expert interviews were particularly relevant, as most knowledge on practical implementation of biosphere reserves resided in the heads of a small number of people who are scattered across different institutions and parts of the world. They range from protected area managers in remote areas to professionals at national governmental nature conservation agencies, and academics at universities and ambassadors and high-ranking officials at UNESCO in capitals and world cities. Access to written information was particularly challenging, as most documents, articles and reports were kept at national ministries or by national authorities. Furthermore, most of the relevant information was only discussed in personal meetings, and little was documented. One advantage for the intended research was that many relevant stakeholders and decision makers in the MAB Programme, especially members of the MAB IAC, work on a voluntary basis or in addition to their normal duties; therefore, they have a particularly high level of personal commitment and interest in the research topic. Thus, it was assumed that they would have a higher willingness to participate in expert interviews and share their knowledge than other respondents. Finally, the final reason for including expert interviews as the main research method for this dissertation was the fact that the implementation of MAB Exit Strategy measures has been quite controversially, especially in countries affected by the Exit Strategy. Therefore, it was assumed that considerable amount of sensitive information was not publicly available in written form but could be requested in personal interviews.

During the research, the selection of experts to assess the impact of the Exit Strategy on the World Network of Biosphere Reserves was mainly influenced by practical factors. In particular, financial and temporal resources were limited; thus, case studies had to be selected from the list of all relevant stakeholders and all 193 member states of UNESCO. First, the existing network of German experts (i.e. delegates to the MAB ICC and MAB working groups and the former head of division at UNESCO and secretary of the MAB Secretariat) were asked for an interview to test and analyse the participation and feedback of the first interviews as well as recommendations of other experts for the interviews. After these first interviews, the research design was considered to be successful and useful; thus, several other experts were named to participate in the study, and contacts were established. In addition, funds were successfully raised to finance trave within Europe and, later, global travel. To answer the research questions and fulfil the global representation of experts, it was particularly important to ensure a geopolitical balance between UN world regions during the selection of expert to cover all relevant institutions and actors in the MAB Programme and to pay special attention to UNESCO member states that were strongly affected by the Exit Strategy. The objective was not the sheer number of interviews but rather to interview as many experts as possible from different member states and institutions. The selection of experts was based on the following criteria:

- Delegates of the MAB ICC and (vice-)chairs of the MAB Bureau
- Members of the MAB IAC
- Member states that voluntarily withdrew biosphere reserves from the World Network after the establishment of the Exit Strategy in 2013
- Geopolitical balance (at least one interview with a representative in each UN region)

Experts were contacted based on recommendations from experts who had already been interviewed, the contact details of national MAB focal points, members of MAB National Committees and lists of participants at MAB ICC meetings.

Based on recommendations from DUNN (2016), most expert interviews were conducted face-to-face or via telephone (DUNN 2016). For this research project, conducting interviews via telephone would have been convenient, as undertaking face-to-face interviews with global experts is very time-consuming and costly. According to CHRISTMANN (2009), telephone interviews are very demanding and difficult. Whether they are useful depends on the type of questions asked in a study. Compared to face-to-face interviews, their disadvantages include a lack of non-verbal cues and potential external disturbances that are usually not recognized. By contrast, face-to-face verbal exchanges can make it easier for the interviewer to identify whether a question was misplaced, and drawn opinions and tentative conclusions can also be checked and verified (pp. 170-172). Therefore, this study focused on establishing as many face-to-face interviews as possible and using the travel budget to schedule them for each of the above mentioned criteria groups. However, additional interviews with key experts were also conducted via telephone or video calls.

When conducting interviews, DUNN (2016) distinguishes between three major forms of interviewing: structured, unstructured and semi-structured. Structured interviews follow a predetermined and standardized list of questions, whereas unstructured forms of interviewing are controlled by the interviewee rather than by a questionnaire. Semi-structures interviews can be seen as a middle course which have some degree of predetermined order but maintain flexibility in the way issues are addressed. The semi-structured interviews employ an interview guide that is organized around ordered but flexible questions. Guidelines are therefore developed to structure the subject area of the study and as a concrete assistance in the survey situation (BOGNER et al., 2014, pp. 27-28).

To address the two main research aims in this study, the questionnaire for the expert interviews focused on the five research topics reflected in the research questions: the current greatest challenges in the implementation of biosphere reserves, programmatic challenges since the Seville Strategy and the adoption of the Statutory Framework, recent quality improvement strategies, interlinkages with other UNESCO labels and the number of biosphere reserves needed in the world (see Annex 2).

To break the ice and begin the interviews with questions that were easy to answer for experts, the following open-ended and broadly formulated questions was asked: 'What are the biggest challenges that UNESCO Biosphere Reserves are facing at the moment? Globally and in your country?' Subsequent questions directly concerned on the five research topics and were more detailed.

The first interview question focused on paradigmatic changes caused by the Seville Strategy and ensuing challenges and problems. Then, questions were asked about the integration of biosphere reserves into national legislation and the use of the term 'biosphere reserve' in the respondent's country. The next six questions concerned quality management, including the current quality management system used in the participant's country, the general attitude of the expert and key actors in their country towards the periodic review system, how it could be improved, the role that regional and thematic networks could play in this process, why the Exit Strategy was necessary and the main challenges in its implementation. The next part of the questionnaire focused on the complementarity or possible clashes between the three UNESCO labels and current challenges with UNESCO labels in the respondent's country. These guiding questions helped the interviewer address the most important topics, keep the flow of the interview relatively comparable in all interviews and provide structure to intensive, face-to-face conversations that lasted around one hour each. Depending on the progression of the interviews, follow-up or supplementary questions were asked to collect further data for the research questions. The questionnaire was only sent out in advance upon the participant's request, with the indication that the questions were flexible and only intended as a guide throughout the interviews. In such cases, it was especially important to ask additional questions to the expert.

At the end of all interviews, a broader question was asked about the mission of biosphere reserves as models for sustainable development and with a view to the future: 'As biosphere reserves are seen as model regions for sustainable development, how many do we need? Are 686 too many or not enough?' The last question was intended to move focus away from difficult questions about the challenges associated with biosphere reserves and the Exit Strategy and to give experts an opportunity to voice their own opinions on the concept and their vision of the future, thereby ending the interviews on a more positive note.

A total of 31 expert interviews were conducted between January and November 2018 (see Table 11). The specified criteria for the selection of experts were successfully met. At least 13 participants are or were delegates to the MAB ICC and/or are or were (vice-)chairs of the MAB Bureau. In addition, the current director and two former directors of UNESCO's Division of Ecological and Earth Sciences and the secretary of the UNESCO MAB Programme were interviewed. Moreover, three members of the MAB IAC were interviewed. Experts in all six UNESCO member states that voluntarily withdrew biosphere reserves from the World Network of Biosphere Reserves between 2013 and 2018 were interviewed. The equal geopolitical distribution of interviews was also achieved; at least one interview was conducted in each UN region. All interviews were recorded with a recording device and saved as mp3 files in a cloud-based database. The interviews were mainly conducted in English; interviews with German-speaking experts were conducted in German, and interviews with Spanish-speaking experts in Spain and Latin America were conducted in Spanish. Collectively, the 31 interviews had a total recording time of 2,297 minutes (around 38.3 hours). On average, each interview lasted around 74 minutes.

Table 11: List of expert interviews conducted for this thesis

	Date	Location	Expert and organization
1	10.01.2018	Bonn, Germany	**Barbara Engels** Managing director of German MAB National Committee, focal point for UNESCO Biosphere Reserves in Germany and department head of international nature conservation at the BfN
2	11.01.2018	Würzburg, Germany	**Dr. Lutz Möller** Deputy secretary-general and head of division of Sustainable Development, Science at German Commission for UNESCO and German delegate to the MAB ICC
3	22.01.2018	Skype	**Dr. Thomas Schaaf** Former director of the UNESCO Division of Ecological and Earth Sciences and secretary of the MAB Programme (2012–2013)

	Date	Location	Expert and organization
4	29.01.2018	Bratislava, Slovakia	**Dr. Julius Oszlanyi** Director of the Institute of Landscape Ecology of the Slovak Academy of Sciences, former chair of the Slovak MAB National Committee and former vice-chair of the MAB ICC
5	30.01.2018	Vienna, Austria	**Dr. Günter Köck** Associate staff member of the Austrian Academy of Sciences, former vice-chair of MAB ICC (2004, 2010 and 2012), member of the Austrian MAB National Committee and Austrian delegate to the MAB ICC
6	09.02.2018	Paris, France	**Miguel Clüsener-Godt** Current director of the UNESCO Division of Ecological and Earth Sciences and secretary of the MAB Programme
7	12.02.2018	Perth, United Kingdom	**Prof. Martin Price** Director of the Centre for Mountain Studies at the University of the Highlands and Islands, vice-chair of UK MAB National Committee and UK delegate to the MAB ICC
8	04.04.2018	Madrid, Spain	**Francisco José Cantos Mengs** Department head of international nature conservation and biosphere reserves at the Spanish Ministry for Agriculture, Fisheries, Food and Environment
9	16.05.2018	Bogor, Indonesia	**Prof. Dr. Ir. Y. Purwanto** Executive Director, Indonesian MAB National Committee
10	24.05.2018	Canberra, Australia	**Prof. Peter Bridgewater** Adjunct Professor at Institute of Applied Ecology, University of Canberra, former director of the UNESCO Division of Ecological Sciences and secretary of MAB Programme (1999–2003) and Australian delegate to the MAB ICC
11	18.06.2018	Denver, United States of America	**Patrick Mangan** Special Assistant at the U.S. National Park Service as coordinator for the U.S. MAB Programme
12	19.06.2018	Washington, D.C., United States of America	**John Dennis** Deputy Chief Scientist, U.S. National Park Service
13	21./22.06.2018	Knoxville & Pittman Center, United States of America	**Tom Gilbert** President of U.S. Biosphere Reserve Association
14	04.07.2018	Nijmegen, The Netherlands	**Prof. Marja Spierenburg** Professor at the Department of Anthropology and Development Studies at the Radboud University Nijmegen and member of the MAB IAC

	Date	Location	Expert and organization
15	17.08.2018	Gabrovo, Bulgaria	**Gencho Iliev** Director Central Balkan National Park, Bulgaria
16	20.08.2018	Sofia, Bulgaria	**Tsvetelina Ivanova** National focal point and secretary of Bulgarian MAB National Committee at the Nature Protection Service Directorate in the Bulgarian Ministry of Environment and Water **Vladimir Vladimirov** Bulgarian Academy of Science and chairman of the Bulgarian MAB National Committee
17	28.08.2018	Montpellier, France	**Dr. Didier Babin** Senior researcher at CIRAD (Agricultural research for development), president of MAB France, French delegate to MAB ICC and former chair of the MAB ICC (2016-2018)
18	04.09.2018	Skype	**Catherine Cibien** Director MAB France
19	21.09.2018	Beijing, Ching	**Prof. Wang Ding** Secretary-general at the Chinese Academy of Science Secretary of the Chinese MAB National Committee **Ma Xuerong** Specialist at the Chinese Academy of Science and the Secretariat of the Chinese MAB National Committee
20	23.09.2018	Seoul, Republic of Korea	**Prof. So-Doon Cho** Professor at the School of Life and Environment, Catholic University of Korea and chair of MAB National Committee of the Republic of Korea
21	28.09.2018	Yokohama, Japan	**Prof. Hiroyuki Matsuda** Professor at the Faculty of Environment and Information Sciences, Yokohama National University and member of the Japanese MAB National Committee **Hari Mata** Staff member at the Japanese National Commission for UNESCO, Japanese Ministry of Education, Culture, Sports, Science and Technology
22	17.10.2018	Paris, France	**Roberto Ramirez Aldana** Ambassador and permanent delegate of Honduras to UNESCO **Carlos Maradiga** Deputy permanent delegate of Honduras to UNESCO and current vice-chair of MAB Bureau
23	18.10.2018	Paris, France	**Luis Armando Soto Boutin** Minister Plenipotentiary and Deputy Permanent Delegate of Colombia to UNESCO
24	22.10.2018	Nairobi, Kenya	**Dr. George Eshiamwata** Deputy director Environment and Biodiversity - Natural Sciences Programme, Kenyan National Commission for UNESCO

	Date	Location	Expert and organization
25	22.10.2018	Nairobi, Kenya	**Dr. Jaro Arero** Director Natural Sciences Programme, Kenyan National Commission for UNESCO
26	14.11.2018	Phone	**Göran Blom** Chairperson of Swedish MAB National Committee and staff of the Swedish Environmental Protection Agency
27	15.11.2018	Accra, Ghana and Skype	**Sheila Ashlong** Chief Programme Officer at Ghana MAB National Secretariat and member of the MAB IAC
28	16.11.2018	Skype	**Toomas Kokovkin** Focal point for UNESCO MAB Programme in Estonia and current vice-chair of the MAB Bureau
29	19.11.2018	Phone	**Dr. Salwa Mansour Abdel Hameed** Chair of Sudanese MAB National Committee, current vice-chair of MAB Bureau and member of the MAB IAC
30	23.11.2018	Skype	**Johanna MacTaggart** National Coordinator for Swedish MAB Programme and rapporteur of MAB Bureau
31	30.11.2018	Moscow, Russian Federation	**Dr. Valery Neronov** Deputy Chair Russian MAB National Committee

Source: Own data and illustration

In addition to the 31 abovementioned expert interviews, three scheduled interviews were cancelled. Two last-minute cancellations occurred in Brisbane, Australia and Pretoria, South Africa, after the researcher travelled to the countries, and one cancellation occurred in Muscat, Oman before the beginning of the journey. At all three appointments, the contacts fell asleep afterwards; thus, they were neither rescheduled nor conducted via telephone. Three additional experts in Egypt, Nigeria and Peru expressed interest in participating but did not respond to attempts to schedule an interview.

The vast majority of interviews (80.6%) took place in person, whilst the rest were conducted via telephone or video call. Experiences with telephone interviews were very mixed. In general, the telephone interviews were shorter and less intensive. A few interviews had to be completed quickly; therefore, it was not possible to establish a good rapport between the expert and interviewer. In addition, there were frequently connection problems, and the quality of the calls was often substandard. During the onsite and face-to-face conversations, a good level of trust could be developed between the researcher and participants, which made it easier to address even the most sensitive and delicate issues. After two face-to-face interviews, additional information was also provided after the recording device was turned off. The seriousness of the study and the importance of the interviews were underscored by the researcher's willingness to the travel, which was appreciated by many participants and led to more participation. During the onsite interviews, considerable literature and documents were also provided to the researcher.

4.2 Qualitative content analysis

Thus far, no independent evaluation method for expert interviews has emerged. However, clear preferences can be identified based on the aim of the interviews. For example, qualitative content analysis is best suited to expert interviews that aim to gather information. In qualitative content analysis, expert knowledge is conceptualized as a collection of information that can correctly represent the world (BOGNER et al., 2014, pp. 71-72).

According to MAYRING (2002), qualitative content analysis can be used to systematically examine expert interviews. The latter are broken down into units and processed one after another. In the content analysis process, important aspects are filtered out from of the collected data using a category system. MAYRING divides qualitative content analysis into three main forms: summary, explication and structuring. Summary means reducing the material to its essential contents through abstraction. In explication, additional information is inserted into the individual text parts through paraphrasing to explain them more clearly. Finally, structuring refers to the structured filtration of the material using predefined categories, which the text material is assigned to (pp. 114-116).

GLÄSER and LAUDEL (2010) developed their own method of extracting and analysing raw data from text materials. This form of qualitative content analysis attempts to systematically reduce the amount of information at an early stage and structure it according to the research questions. The core of this process is extraction (i.e. the retrieval of required information from the text). Extraction means reading the text and deciding which information is relevant for the investigation. The process is divided into four main steps: preparation for extraction, extraction, processing of the data and evaluation. During preparation, a category system is defined for a search grid of relevant information, which is based on the preliminary theoretic considerations. However, existing categories can be edited if the text contains information that appears relevant. During extraction, the required information is taken from the search grid and assigned to the appropriate categories. Next, the raw data is summarized, checked for redundancies and contradictions and sorted according to criteria that are relevant to the evaluation. This process results in a structured information base (pp. 200-202).

In this dissertation the analysis of expert interviews was based on GLÄSER and LAUDEL's (2010) the method. First, the parts of the interviews that were relevant for answering the research questions were transcribed in text form using the computer software Express Scribe. The exact wording of the interviews was transcribed, except for slips of the tongue, pauses in conversations and expletives, which were only transcribed in exceptional cases. The transcription of the 31 interviews resulted in a total of 317 written pages; each interview averaged around 10 pages of text.

Second, the categories were defined based on the research topics and categories from the interview questionnaire. A category was created for each research topic and questionnaire question, as well as for all countries considered as case studies in this dissertation. The analysis software Atlas.ti was used to code the transcribed text from all interviews and assign relevant text passages to the corresponding categories. Multiple classifications in categories were also possible. All coded passages were also assigned references (e.g. 2:8) for further verification of the raw data during the analysis. The first number indicated the interview number, and the second number represented the consecutive number of codes in the interview. The numbers for the interviews range up to 35 and refer to the automatic count of the Atlas.ti software, which also counted four additional uploaded documents. It is important to note that after Chapter 6 the numbering of the interviews did not correspond to the chronological order in which they were conducted (see Table 11); instead, a random order using an online random number generator was applied. This measure was taken to avoid directly assigning references to the interviewed experts and to ensure a certain level of anonymity, especially given the sensitive topics in chapters 6 and 7 of this thesis.

Then, the information obtained from the extraction was translated, summarized and sorted according to the relevance of the main statements in Excel. Thus, the text material was reduced to core statements and structured according to content. Based on this structured information base, relevant content was incorporated into Chapters 5, 6 and 7.

5 Case Studies: the complexity of the biosphere reserve concept in practice

To conduct a global assessment of the effects of recently introduced quality enhancement strategies on the World Network of Biosphere Reserves, this study presents differences and similarities in the implementation and quality management of biosphere reserves through country-specific case studies. In addition to a pure analysis of the conceptual framework, it is also important to highlight the individuality of the biosphere reserves. Cultural, social, natural, political and economic conditions widely varied across the world network, and this individuality is precisely embedded in one of the MAB Programme's main aims: to develop model regions for sustainable development and to reflect differences and local challenges. In addition, it is important to note that, although biosphere reserves are internationally recognized by UNESCO, at least their core zones are protected areas under national law and their management falls under the purview of individual states.

Therefore, the following sections provide insights on the challenges related to the implementation of the UNESCO MAB Programme and biosphere reserves in member states and individual biosphere reserves to showcase the complexity and diversity of the concept in practice. More specifically, detailed case studies of 20 UNESCO member states in all five UN regions are presented according to the selection criteria for the expert interviews. The case studies mainly include national and site-specific information gathered from the interviews to ensure comparability and are supplemented by secondary literature or reports where appropriate. The information is mainly derived from the expert interviews and is reproduced as closely as possible to the original. As a result, the writing style is less academic in these case studies and is strongly oriented towards the oral language and the interview notes with the interviewees.

The case studies follow a similar structure. First, they include an overview map of the country in question, which shows the location all biosphere reserves and some basic information. Then, they provide information on the development of the MAB Programme and the biosphere reserves in the given country and the setup of national MAB structures. Next, they describe impacts and challenges related to the implementation of the Exit Strategy. Each section concludes a description of current main challenges and the outlook for the country's biosphere reserve. The case studies are divided by UN region and presented in alphabetical order.

5.1 Africa

5.1.1 Ghana

Figure 12: UNESCO Biosphere Reserves in Ghana

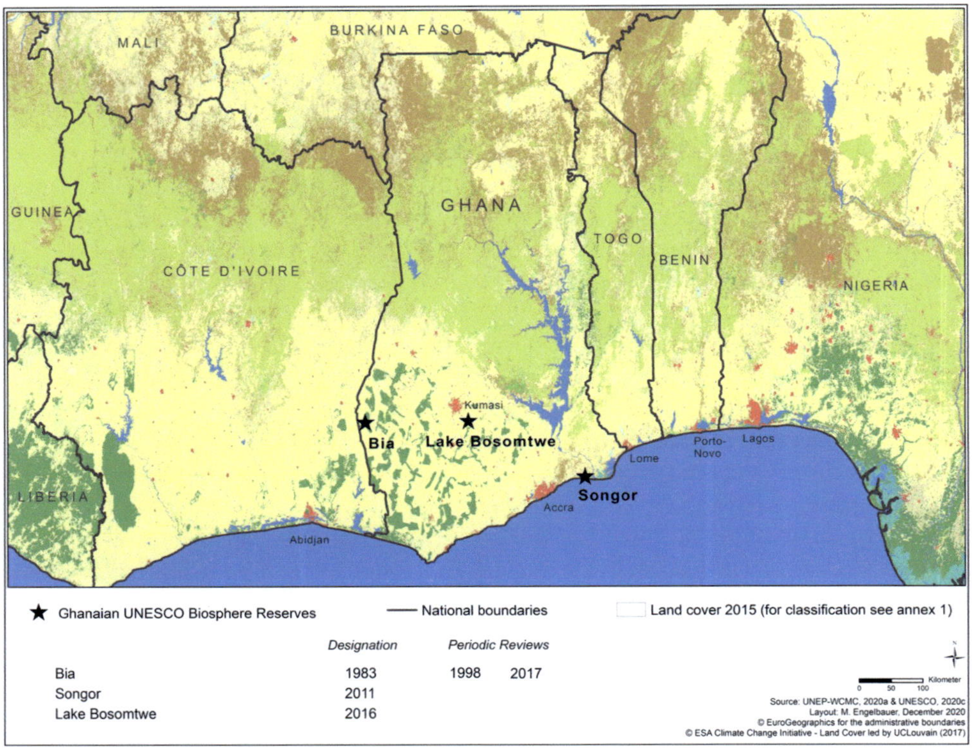

Source: Own illustration

Ghana currently hosts three UNESCO Biosphere Reserves (see Figure 12). The Bia Biosphere Reserve in Ghana; it was designated in 1983 and is in the Western Region, near the border with Côte d'Ivoire. Its moist evergreen and semi-deciduous forest was mostly destroyed through intensive logging and cocoa farming before the establishment of the biosphere reserve (SALU, 2013, pp. 73,77). The biosphere reserve consists of the Bia National Park (core zone), the Bia Resource Reserve (buffer zone) and an estimated 43 neighbouring communities and two forest reserves in the transition zone. Several development projects in the biosphere reserve have promoted sustainable income alternatives to reduce pressure on resources and strengthen ownership of communities in the management of natural resources (ASHONG et al., 2013, p. 125).

In 2011, Ghana's first coastal biosphere reserve was established: the Songor Biosphere Reserve. The second-largest Ramsar site in Ghana is in the Greater Accra Region and encompasses a unique complex of diverse habitats, species and ecosystems of high economic, cultural and biological value. The biosphere reserve is completely owned by communities, with all resources owned by clans. Elders serve as custodians who sell or lease land. Communities highly depend on resources in many ways and are mainly involved in subsistence crop farming, fishing, hunting, salt mining and fuel wood collection. In addition, the biosphere reserve's mangroves and beaches attract many national and international visitors every year, and the area is considered one of the top national tourist destinations. Therefore, sustainable tourism is one of the main economic drivers for local populations (Ashong et al., 2013, pp. 123, 126-127).

Lake Bosomtwe Biosphere Reserve combines forest and wetland ecosystems around one of only few meteoritic lakes in the world; it was listed in the World Network of Biosphere Reserves in 2016. The lake is the result of meteoritic action around 1 million years ago, when a crater formed by an asteroid impact gradually filled with rainwater. Approximately 50,000 people live in the Lake Bosomtwe area; their main economic activities are farming, fishing and tourism (UNESCO, 2017d).

The management system and legal structures of the three biosphere reserves in Ghana differ, as there is no specific legislation for biosphere reserves. At Bia Biosphere Reserve, the core zone is designated as a national park under Ghanaian legislation; however, the buffer and transition zone focus more about consulting and educating people about the benefits of the biosphere reserve (29:2). The second biosphere reserve, Songor, was already a Ramsar site at the time of its designation and featured a highly active management committee consisting of the forestry commission, local communities and other stakeholders. When it became a biosphere reserve, the communities became involved in activities; one traditional leader of a community is also the chairman of the biosphere reserve and is involved in its day-to-day management. Lake Bosomtwe Biosphere Reserve is a unique case, as it was neither a protected area nor protected by any national legislation. With its designation as a biosphere reserve, a new kind of legal system was developed with communities and within the laws of the forestry commission. A designated area for protection and a management committee was created, which is mainly overseen by community members (29:3).

Ghana's MAB National Committee currently has 28 members who represent government institutions, research institutions, universities and other organizations whose activities are related to natural resource management and conservation. The committee is hosted by the Environmental Protection Agency and works closely with the two management authorities for biosphere reserves in Ghana: the Forestry Commission and the Water Resources Commission (Ghana MAB National Committee, 2018, p. 1). There have long been discussions about establishing a specific law for biosphere reserves in Ghana; however, there is still debate about how to divide responsibilities between the two government agencies that oversee the MAB Programme in Ghana (29:4; 29:5).

In November 2018, the first National Forum on Biosphere Reserves was held in Accra. It gathered many stakeholders from across Ghana to discuss the relevance of UNESCO Biosphere Reserves to meet sustainable development goals and the role of various stakeholders. During the national forum, the Ghana Action Plan, a national version in contextualization of the Lima Action Plan, was presented (see Figure 13; Ghana MAB National Committee, 2019, p. 2).

Figure 13: First National Forum on UNESCO Biosphere Reserves in Ghana

Source: M. Engelbauer, November 2018

Bia Biosphere Reserve, which received its designation during the pre-Sevilla phase, was affected by the Exit Strategy, as it was as a typical national park designated as biosphere reserve. Thus, Ghana had to make changes to the biosphere reserve to meet the new objectives established in the Seville Strategy. After numerous meetings and conversations with stakeholders, the biosphere reserve was rezoned, and a new nomination was successfully submitted to and accepted by the MAB ICC (29:1).

Ghana has been very active in the creation and development of the legal framework for AfriMAB and chaired the sub-network for two consecutive terms before handing over control to Nigeria. Regional integration is an important tool that provides people and biosphere reserve managers opportunities to learn from other countries, as for biosphere reserve managers it is essential to learn that some challenges in the implementation of the concept and activities are not unique to one country only (29:9).

So far, there are neither natural World Heritage Sites nor UNESCO Global Geoparks in Ghana. The only World Heritage Sites are cultural ones, consisting of coastal forts and castles and traditional Asante buildings. In general, there is a very low level of awareness about different UNESCO labels among the society. Even Biosphere Reserves are better-known than World Heritage Sites but so far, the point is not reached where the people are conceiving the label yet (29:13).

In Ghana, the future of the MAB Programme is less about the number of designated biosphere reserves and more about ecosystems. The plan is to represent every ecosystem. So far, a forest, wetlands and a natural lake have been designated as biosphere reserves, which means that drylands and a mountain area are still missing (29:14). The MAB National

Committee has earmarked 16 sites for future nomination as biosphere reserves, including several initiatives for transboundary biosphere reserves (GHANA MAB NATIONAL COMMITTEE, 2018, p. 5). One stakeholder consultation was initiated about a mountain range that stretches from Ghana to Togo; there have been ongoing discussions to nominate a transboundary biosphere reserve around Kyabobo National Park since 2015. This seems quite possible, and the current idea is to designate it in Ghana first. Another potential transboundary biosphere reserve can be planned together with Burkina Faso; both governments already signed a Memorandum of Understanding to create a transboundary site around Gbele Resource Reserve, and all consulted stakeholders were receptive to the idea. It is very likely to be nominated within the next three years (29:15; GHANA NATIONAL COMMISSION FOR UNESCO, 2019, p. 14; GHANA MAB NATIONAL COMMITTEE, 2019, p. 5).

5.1.2 Kenya

Figure 14: UNESCO Biosphere Reserves in Kenya

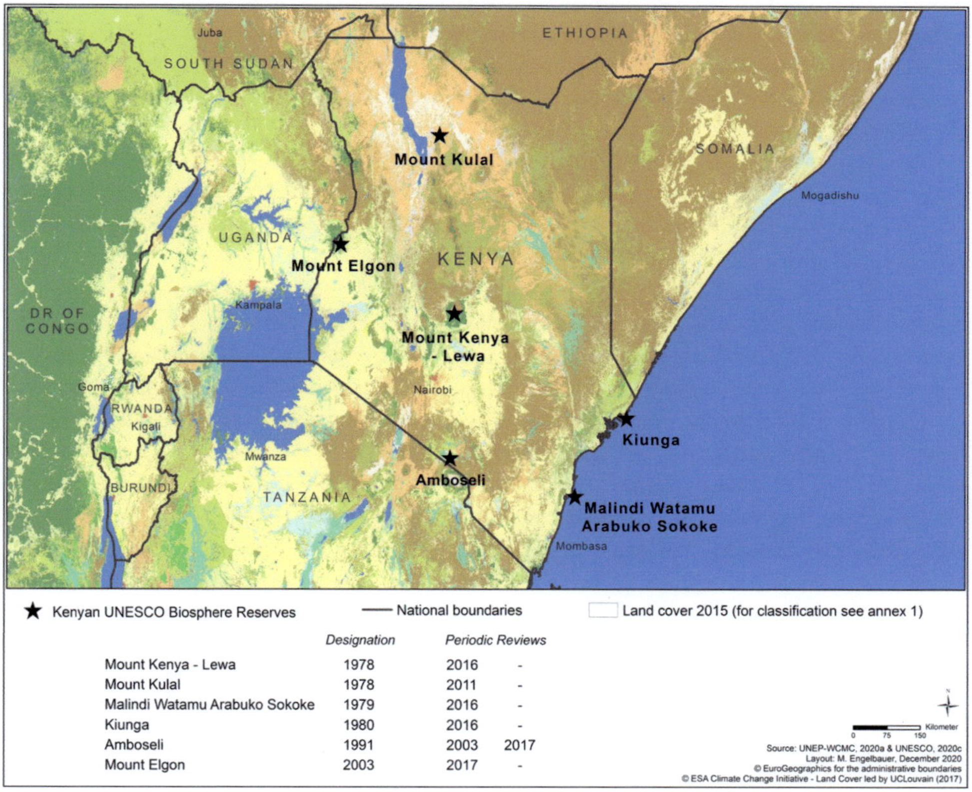

	Designation	Periodic Reviews	
Mount Kenya - Lewa	1978	2016	-
Mount Kulal	1978	2011	-
Malindi Watamu Arabuko Sokoke	1979	2016	-
Kiunga	1980	2016	-
Amboseli	1991	2003	2017
Mount Elgon	2003	2017	-

Source: Own illustration

In Kenya, there are currently six UNESCO Biosphere Reserves (see Figure 14). Three of these are mountain and highland biosphere reserves: Mount Kenya Lewa Biosphere

Reserve, which includes the second-highest mountain in Africa and the highest in Kenya; Mount Elgon Biosphere Reserve on the border with Uganda; and Mount Kulal Biosphere Reserve in northern Kenya, which surrounds an eroded and extinct volcano at the south-eastern edge of Lake Turkana – one of the driest regions in East Africa. In addition, Kenya contains two marine and coastal biosphere reserves: Kiunga Biosphere Reserve and the recently extended Malindi Watamu Arabuko Sokoke Biosphere Reserve. Finally, the last site is Amboseli Biosphere Reserve at the foot of Mount Kilimanjaro on the border between Kenya and Tanzania; it includes a variety of ecological zones, from natural dry mountain forests to savannah rangelands, wetlands and swamps (UNESCO, 2020c).

The MAB Programme in Kenya is managed by the Kenyan National Commission for UNESCO (NATCOM Kenya). Its main tasks are to support biosphere reserves and host the national MAB Committee. The MAB National Committee is a multi-agency committee consisting of, amongst others, the Kenya Wildlife Service (management of protected areas and national parks), the Kenyan Forest Service (management of forest ecosystems), the National Environment Management Authority (development of policies for ecosystem and biodiversity conservation) and the National Commission for Science and Technology (responsible for advancing science, technology and innovation). Other members of the MAB National Committee include academics at Kenyan universities and research institute; they aim to link academic research and conservation, involve students in research in the biosphere reserves and ensure that research aligns with the sites' priorities. The former chair of the committee was a professor at one of these universities. NATCOM Kenya also hosts the Secretariat of the MAB National Committee and established an expert committee to ensure the implementation of activities for conservation, raise awareness and support development at designated biosphere reserves (24:1; 24:8). The periodic review is developed by the MAB National Committee and the national MAB Secretariat (24:13).

UNESCO Biosphere Reserves in Kenya face several challenges. In addition to site-specific challenges, such as human encroachment, population increase, desertification and the overexploitation of resources, especially in buffer and transition zones, the main challenge is funding. The sustainable management of the sites and collaboration with communities to ensure conservation, protect their livelihoods and engage in the necessary logistics, requires significant funding and intensive coordination between all stakeholders, which is another challenge (24:2). For example, if all five institutions that are involved in the management of the biosphere reserves would pool and mobilize their resources, funding would not be an issue. However, these institutions have different priorities. Thus, the work of the MAB National Committee is crucial in balancing different needs, enabling closer cooperation and linking all institutions to the biosphere reserves (24:3; 24:7).

In recent years, human-wildlife conflicts have increasingly posed a challenge for Kenyan biosphere reserves. The main reason is that people have occupied areas that were initially intended as wildlife corridors. As a result, they lose their livestock or crops, then take revenge by killing wildlife. They feel neglected because the government does not intervene or support them. At least three Kenyan biosphere reserves experienced similar problems and faced the challenge of how to maintain wildlife corridors whilst protecting the interests and preventing the resentment of local communities (24:4).

Regarding the designation of biosphere reserves, it is also important to acknowledge political changes and changes in land tenure of the initial boundaries of the sites. There was a need to redefine zones or extend certain sites for the purposes of the MAB Programme and the Statutory Framework. For example, according to the dossier for the Mount Kulal Biosphere Reserve in 1978, the National Environmental Secretariat was responsible for the site's management; however, more than 50% of the area is now managed by communities. It is a prime example of ownership by local communities, as they have taken responsibility for the biosphere reserve's management and consider it to be crucial (24:9).

However, management systems differ for each biosphere reserve in Kenya. For example, Mount Kenya Lewa Biosphere Reserve is under the jurisdiction of both the Kenya Wildlife Service, which manages the national park in the core zone, and the Kenya Forest Service, which oversees the buffer and transition zone. Thus, responsibility for the management of the biosphere reserve lies with both institutions, but there is no dedicated individual manager who is responsible for the entire biosphere reserve. It is the responsibility of NATCOM Kenya and the Kenyan MAB Secretariat to offer technical support and to ensure that both institutions work together (24:15). Five Kenyan biosphere reserves are also national parks and therefore managed in part by the Kenya Wildlife Service. The sixth site, Mount Kulal Biosphere Reserve, is managed by local communities as a conservancy, without any government involvement. One advantage of this approach is that local communities treat the core zone as a sacred cultural site; thus, it is of religious and cultural importance to them. However, their cultural connection to the land is changing with the change in generations from their grandparents and parents; thus, local organizations are also changing. As a result, changes in perception could endanger the biosphere reserve, which demonstrates the need for more frequent evaluation and review (26:3).

Although five out of six biosphere reserves in Kenya are first-generation biosphere reserves, only two were affected by the Exit Strategy. The first one, Malindi-Watamu (named at the time), had to add a new terrestrial core zone to achieve the required zonation, which resulted in a completely new biosphere reserve submission. The main challenge was securing the necessary funding to tackle the rezonation and submission to the MAB Secretariat, which considerably delayed the process. However, the application was successfully submitted and approved with close collaboration from partners, and the biosphere reserve is no longer affected by the Exit Strategy (26:1). The second affected biosphere reserve was Kiunga, which is near the border with Somalia. The security situation in this area changed in recent years, and it is now a no-go-zone; not even the Kenyan Defense Force is allowed to go there. Therefore, it was challenging to undertake the rezonation of the biosphere reserve under these circumstances (26:4).

In addition, there is currently a paradigm shift taking place in nature conservation in Kenya. The days of pure protectionism are over, as this leaves no options to work with communities; it is important to provide some benefits to the people, and the only way to do so is by working with them. Thus, in Kenya, the biosphere reserves model is very attractive for addressing human populations' increasing pressure on protected areas, as it includes and considers people in the utilization of natural resources but limits it to certain zones. Certain areas are set aside for conservation and others for both conservation and the utilization of resources. Thus, biosphere reserves provide one solution to for many challenges, such as how to ensure the sustainable utilization of resources and collaboration with communities and how to provide some benefits both communities and the environment (24:18).

Over the next few years, there are three main initiatives for new biosphere reserves. One is situated in the desert, next to a mountainous area in which desertification is rapidly occurring; this is a very critical ecosystem. The area would be an ideal biosphere reserve, as the core zone is already protected by the Kenya Wildlife Service and surrounded by a game reserve, which could serve as a buffer zone. Moreover, several communities live around the area and rely on ecosystem services in the core and buffer zones. The second initiative is a possible transboundary biosphere reserve at Mount Elgon that would be jointly overseen with Uganda, and the third one is an even larger initiative for a transboundary biosphere reserve involving seven countries around Lake Victoria. Although only 6% of Lake Victoria is located in Kenya, this area corresponds to the most polluted part of the lake (26:6; 24:6).

Important aspects for future work on Kenyan biosphere reserves are branding and awareness raising. People who live in biosphere reserves must know what it means to be a biosphere reserve. Awareness raising at the community level, such as at schools and churches, is needed to ensure that they know what biosphere reserves are and how they can benefit from them. For example, the locally produced goods such as tea and coffee can be branded (24:11). In addition, the management of multiple designations must be examined in depth to ensure that they are synergistic and complementary (26:5).

5.2 Arab states: Sudan

Sudan has two first-generation biosphere reserves that were designated in 1979 and are managed as national parks and one second-generation biosphere reserve designed that was designated in 2017 (see Figure 15).

The first site, Dinder Biosphere Reserve, includes one of the oldest protected areas in Africa. The national park in the core zone was established in 1935 following the London Convention on the conservation of African flora and fauna and is situated in the southeast part of the country, near the Ethiopian border. In addition to the national park, the biosphere reserve also covers the Rahad Game Reserve in the north. The area consists of low-lying floodplains that slope gently from the Ethiopian highlands and are the origin of the Rahad and Dinder rivers (ROBERTSON, 2001, p. 885). Human settlements in the region date back at least several centuries and are characterized by several migration cycles. Since the early 1960s, settlement in the area has intensified due to the immigration from western Sudan and West African countries, which resulted from famine and drought (VAN HOVEN & BASHIR NIMIR, 2004, p. 28). Large parts of the buffer zone have been degraded by mechanized farming and woodcutting for charcoal production (ROBERTSON, 2001, p. 885).

The greatest challenge for the Dinder Biosphere Reserve is that many members of local communities illegally enter the site to extract natural resources and sell them in other places for their livelihoods. Most local communities that surround the biosphere reserve are poor and depend on the use of natural resources. Thus, it is important for site managers to hold meetings with stakeholders to explain the consequences of natural resource loss and attempt to change their opinions about biosphere reserves managers (31:3). The Sudanese MAB

National Committee is currently working on establishing a transboundary biosphere reserve with the Ethiopian MAB National Committee. So far, the progress is promising, but the biggest challenge remains financial support (31:7).

Figure 15: UNESCO Biosphere Reserves in Sudan

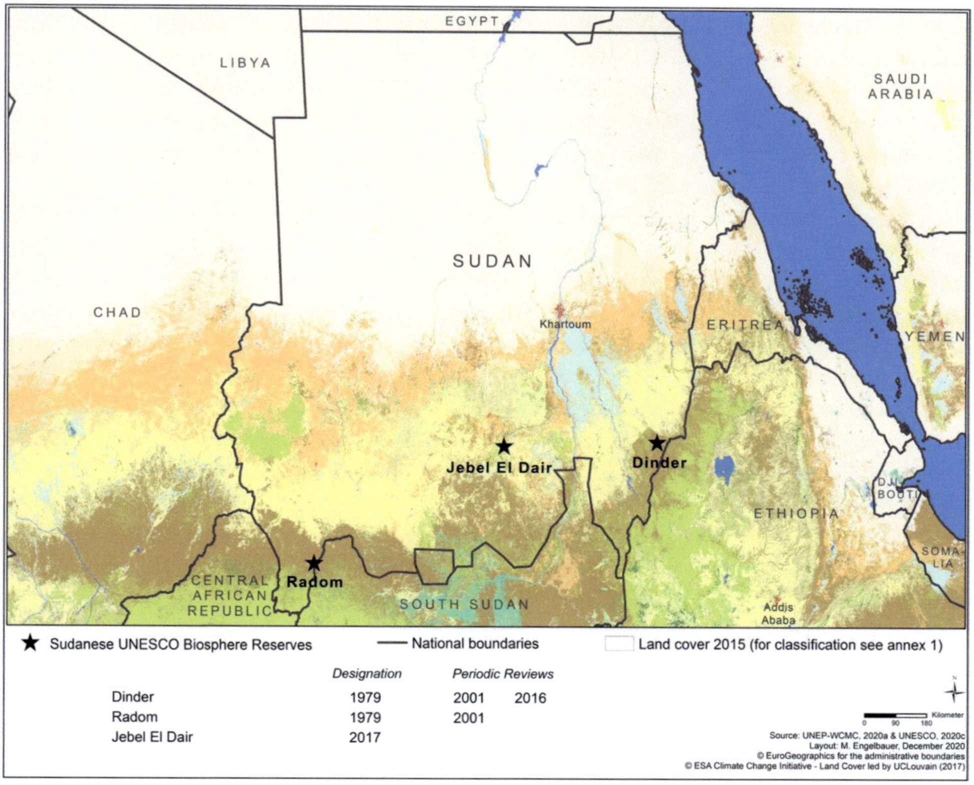

Source: Own illustration

The second site, Radom Biosphere Reserve, was designated in 1979 and is located in the southwestern part of the country; it directly borders South Sudan and the Central African Republic. Most of the biosphere reserve is disputed between Sudan and South Sudan. Because it is in a remote conflict area, it is difficult for managers to visit the site due to the security situation; for researchers, it is impossible. The area consists of broken hilly country with rivers, streams and permanent pools and is characterized by a very rich woodland savanna area. Some trees can only be found in this part of Sudan (31:6; ABDEL HAMEED, 1998, p. 23). It is believed that the site has suffered from commercial game poaching, and the possible decline of tsetse fly populations allowed the ingress of domestic herds, which resulted in overgrazing (ROBERTSON, 2011, p. 886). The greatest challenge in the management of the biosphere reserve is its remoteness and the security situation. To hold meetings in the biosphere reserve, cars, security forces and petrol are needed; however, the budget is very limited because of the economic situation in Sudan and the government's other priorities.

Whilst reports indicate that biodiversity remains intact, it is crucial to visit the site and assess the situation (31:8).

Finally, Jebel Al Dair Biosphere Reserve was designated as a biosphere reserve with the active involvement of surrounding local communities, which are also part of the management system (31:2). The site is part of the Al Dair mountain range and consists of dry savannah woodlands, forest ecosystems and a network of streams. The rich biodiversity of the semi-arid region remains intact and is valued by local stakeholders as an economic source of livelihood. A socio-economic study conducted in 2012, found that local inhabitants benefited from the biosphere reserve through wood collection (28%); grazing (26%); the collection of medicinal, aromatic plants and grasses (23%); the collection of fruits and honey (12%); and bush meat hunting (9%; UNESCO, 2020c).

The biggest challenge for the Sudanese biosphere reserves is the acceptance and participation of local communities. This is particularly the case with the Dinder and Radom Biosphere Reserves, as both were originally designated as national parks and many local people still see them as purely protected areas. However, it is also true for managers, as both biosphere reserves are still managed as national parks and their directors are old-fashioned conservationists. Thus, significant training is needed on how to implement the concept and engage and earn the respect of local communities (31:1). A new project funded by UNEP that focuses on the biodiversity of protected areas aims to begin raising awareness amongst poor populations about the benefits of nature and biodiversity and how they can support their livelihoods (31:4).

Moreover, it is also important to persuade the government to increase investment in the biosphere reserves, because it has mainly focused on large cities rather than remote areas thus far (31:4). When government and ministers change, they must be convinced anew of the importance of biosphere reserves (31:9).

The MAB National Committee in Sudan includes people from research institutes and universities, managers of biosphere reserves, staff from the Forestry Department and Ministry of Education and members of wildlife and nature conservation NGOs. In the future, there are also plans to include representatives from the private sector. The committee meets two or three times per year and is overseen by the UNESCO National Commission (31:11). At the national level, several lectures are held at universities to introduce students to the MAB Programme and the Sudanese biosphere reserves. The minister was also asked to nominate one of these youths to the MAB National Committee (31:12).

The authority that is responsible for managing the MAB Programme and the biosphere reserves in Sudan is the Wildlife Conservation Administration. The periodic reviews are conducted by the directors of the biosphere reserves with support from the MAB National Committee, then submitted to the UNESCO National Commission (31:13). The greatest challenge in the preparation of the periodic reviews is communication with the directors of biosphere reserves, as they are in remote locations and there is no internet connection at the sites. They can only be contacted via their personal mobile phones; sometimes university researchers can take documents to the biosphere reserves when they travel to the areas (31:14).

Radom Biosphere Reserve was affected by the Exit Strategy due to the need to rezone the site. However, as it is still part of an armed conflict, the decision was made to keep the biosphere reserve in the World Network of Biosphere Reserves until the conflict is resolved (31:16).

5.3 Asia and the Pacific

5.3.1 Australia

Figure 16: UNESCO Biosphere Reserves in Australia

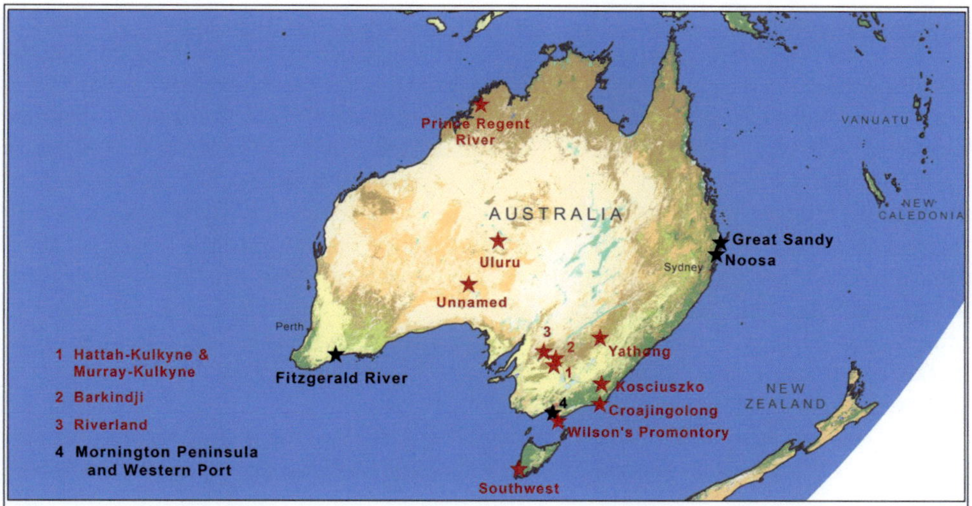

★ Australian UNESCO Biosphere Reserves — National boundaries ▭ Land cover 2015 (for classification see annex 1)

★ Withdrawn UNESCO Biosphere Reserves

	Designation	Periodic Reviews		Withdrawal
Croajingolong	1977	2003		2020
Kosciuszko	1977	2003		2020
Southwest	1977			2002
Macquarie Island *	1977			2011
Prince Regent River	1977			2018
Unnamed	1977	2003		2020
Uluru	1977	2003		2020
Yathong	1977	2003		2018
Fitzgerald River	1977	2003	2017	
Danggali **	1977	2004		2020
Hattah-Kulkyne & Murray-Kulkyne	1981			2018
Wilson's Promontory	1981			2018
Morning Peninsula and Western Port	2002	2015		
Barkindji	2005			2018
Noosa	2007	2019		
Great Sandy	2009			

* not shown on the map above
**extended and renamed Bookmark in 1995; renamed again as Riverland in 2004

0 300 600 Kilometer

Source: UNEP-WCMC, 2020b & UNESCO, 2020c
Layout: M. Engelbauer, December 2020
© EuroGeographics for the administrative boundaries
© ESA Climate Change Initiative - Land Cover led by UCLouvain (2017)

Source: Own illustration

Australia is a federal country; when the former colonies agreed to form the Commonwealth of Australia, they insisted that federal states had full responsibility for the environment. In 1901, decisionmakers were not aware of the possible consequences; however, the current interpretation is that states have full responsibility for all land, water, air and wildlife within their jurisdiction. The only exception for the federal government is, what is called, foreign affairs power. When Australia signs a convention or an agreement, such as the World Heritage Convention, then the responsibility lies with the federal government (11:7). Since the MAB Programme is a programme and not an agreement or a convention, biosphere reserves do not fall under federal jurisdiction. Whilst all states have legislation that addresses protected areas, none has legislation that deals with biosphere reserves. Consequently, biosphere reserves are not governed by any Australian legislation; only their core areas are protected areas that fall under state responsibility and legalisation (11:8).

When the MAB Programme was initiated in 1972, Australia underwent a complete change of government, switching to a very radical labour government after 23 years of liberal rule. The new government signed the World Heritage Convention and established the National Parks and Wildlife Service (11:9). The latter was approached by the UNESCO National Commission in 1975 to participate in the new MAB Programme and to nominate biosphere reserves. Between 1976 and 1982, 12 Australian biosphere reserves were added to the World Network of Biosphere Reserves (see Figure 16); however, they were little more than nature reserves or national parks. Like many first-generation biosphere reserves, they focused on the conservation function and received a certificate, but little happened afterwards. The situation largely remained the same until the Seville conference (11:10). According to MATYSEK et al. (2006), a certain lack of capacity or unwillingness to contribute to international improvements to the MAB Programme meant that there was little information about Australia's biosphere reserves; thus, the biosphere reserves were not taken very seriously in policy matters (p. 89).

The Seville Strategy and Statutory Framework showed that the Australian biosphere reserves did not fit the developed criteria, except for the Danggali and Fitzgerald Biosphere Reserves. Adjacent to the Danggali biosphere reserve, the Calperum Pastoral Station was ubicated, which had become overgrazed and severely degraded. The federal government was convinced that the biosphere reserve approach would rescue the station and bought it with financial support of the Chicago Zoological Society. The Danggali Biosphere Reserve was subsequently extended, re-nominated as Bookmark Biosphere Reserve and approved by the MAB ICC in 1994. The state and federal governments gave ownership and responsibility for the management of the entire regional landscape to local communities; several initiatives, such as harvesting goats for meat and skins, were implemented. In the late 1990s, Australian state governments saw Bookmark Biosphere Reserve as a model for renewing existing biosphere reserves and creating new ones, as it had attracted significant philanthropic, governmental and NGO investment (BRIDGEWATER, 2020, pp. 232-233; BRUNCKHORST, 2001, pp. 26-27). In 2004, biosphere reserve managers completed the periodic review and renamed the site to Riverland Biosphere Reserve. However, the complexity of management bodies and waning interest from governments resulted in the second periodic review being overdue for several years. Ultimately, the biosphere reserve was withdrawn from the

World Network of Biosphere Reserves in 2020, as it could not meet its original high expectations (BRIDGEWATER, 2020, pp. 233-234).

One positive example of a first-generation biosphere reserves is the Fitzgerald Biosphere Reserve in Southwest Australia. The community adjacent the Fitzgerald national park is a very active cultural place without a larger urban area. It became interested in extending the biosphere reserve and rejuvenating their farmlands through innovative conservation ideas. After a long process, the biosphere reserve was extended, rezoned and re-nominated in 2017 with a strong community part next to the West Australian government involvement for the management of the national park as a core zone. Now this biosphere reserve contains an ideal balance between the state government and the local community (11:14).

At the beginning of the 21st century, the local community of Mornington Peninsula was interested in creating a biosphere reserve on the outskirts of Melbourne with financial support from the Chicago Zoo and other philanthropic foundations. The biosphere reserve spans two bays in the southwest of Melbourne and includes Philipp Island, French Island and the Western Port Bay Ramsar site (11:12). The core area comprises approximately 4% of this site; it is surrounded by a primarily marine buffer zone, which comprises 30% of the site, and the remaining transition zone has a permanent population of around 300,000 people, with an additional 60,000 people during the summer vacation period (MERCER & HYMAN, 2009, p. 413). The Mornington Peninsula and Western Port Biosphere Reserve was approved by the MAB ICC in 2002 and is the first Australian biosphere reserve driven by local communities. Since its designation it has its ups and downs, as the local council that manages the site has limited financial resources (11:12).

In 2006, the former chair of the UNESCO National Commission and residents of Noosa, Queensland were excited about the idea of a biosphere reserve, as Noosa was a very special place with small towns, forests, national parks and rural land holdings. In addition, the local council was the first one in Australia to introduce an environment levy. In 2007, the Noosa Biosphere Reserve was approved by the MAB ICC, like the Mornington Peninsula and Western Port Biosphere Reserve, was by the local community with almost no involvement from the government. Two years later, a local community a bit north of Noosa nominated the Great Sandy Biosphere Reserve, which consisted of the Great Sandy Strait Ramsar site and the Fraser Island natural World Heritage Site. Both biosphere reserves remain very active and organized the first Australian Biosphere Conference in 2019 (11:13).

Whilst these three community-driven sites enriched the Australian network of biosphere reserves, others became a concern. For example, the Southwest National Park in Tasmania was not aligned with the Seville Strategy and was withdrawn from the World Network of Biosphere Reserves in 2002. In 2011, after reviewing the future activities of the sub-Antarctic Macquarie Island Biosphere Reserve, the Australian authorities realized that it could not be applicable to the Statutory Framework and delisted the site. With the adoption of the Exit Strategy, the focus clearly turned to the future of Australia's first-generation biosphere reserves. Apart from the Fitzgerald Biosphere Reserve, the remaining eight first-generation biosphere reserves and the non-functioning Barkindji Biosphere Reserve were simply incompatible with the Statutory Framework. Thus, the Yathong, Prince Regent River, Hattah-Kulkyne and Murray Kulkyne, Wilson's Promontory and Barkindji Biosphere Reserves were withdrawn in 2018 and Uluru, Kosciuszko, Unnamed and

Croajingolong Biosphere Reserves were withdrawn in 2020 (Bridgewater, 2020, pp. 236-237; 11:15).

In Australia the use of the term 'reserve', under Australian state legislation refers to strict protected areas, without public access. Since this contradicts the concept of biosphere reserves, they do not have the word 'reserve' in their name in Australia. Instead, they are simply called Fitzgerald Biosphere or Noosa Biosphere, for example (11:2).

In recent years, Australia has aligned its 15 biosphere reserves with the Statutory Framework and cleaned up the decks accordingly. After several rezonation attempts, only four post-Seville biosphere reserves remained. At the time of the survey in 2018, there were no ongoing plans for new designations in Australia, as this was not a national government priority. In the future, it will be important to use the biosphere reserve model to assist indigenous populations to manage the lands through an international recognition, which can be an economical asset to them (11:37).

5.3.2 China

After Chinese scientists participated in the UNESCO conference in 1971, they suggested the establishment of a national MAB committee. In 1978, the State Council of China approved the establishment of a Chinese National Committee for UNESCO's MAB Programme (commonly known as China MAB) and established the China MAB Secretariat at the Chinese Academy of Sciences (19:1). The first three UNESCO Biosphere Reserves were designated in 1979: Changbaishan, Dinghushan and Wolong Biosphere Reserves. Since then, many more biosphere reserves have been established in China; as of 2020, there are 34 UNESCO Biosphere Reserves in the country (see Figure 17). After Spain, the Russian Federation and Mexico, China has the fourth most biosphere reserves in the world. Most China's biosphere reserves (25 out of 34) are located in mountain and forest ecosystems; however, China also has two island sites, two coastal sites, two freshwater lake sites and two grassland and steppe sites. All of these biosphere reserves are of high conservation value and enable local communities to continue their traditional ways of life in transition areas (Ding & Ounli, 2020, p. 214).

The Chinese Academy of Sciences provides financial and technical support to the 10-member Chinese MAB Secretariat and the MAB National Committee (19:3). The biosphere reserves are operated by a management committee and receive their budget from the relevant managing state department (19:10). The management committee consist of the bureau for the protected area, the local government and local communities and stakeholders. Together, they discuss the action and management plan (19:12).

Figure 17: UNESCO Biosphere Reserves in China

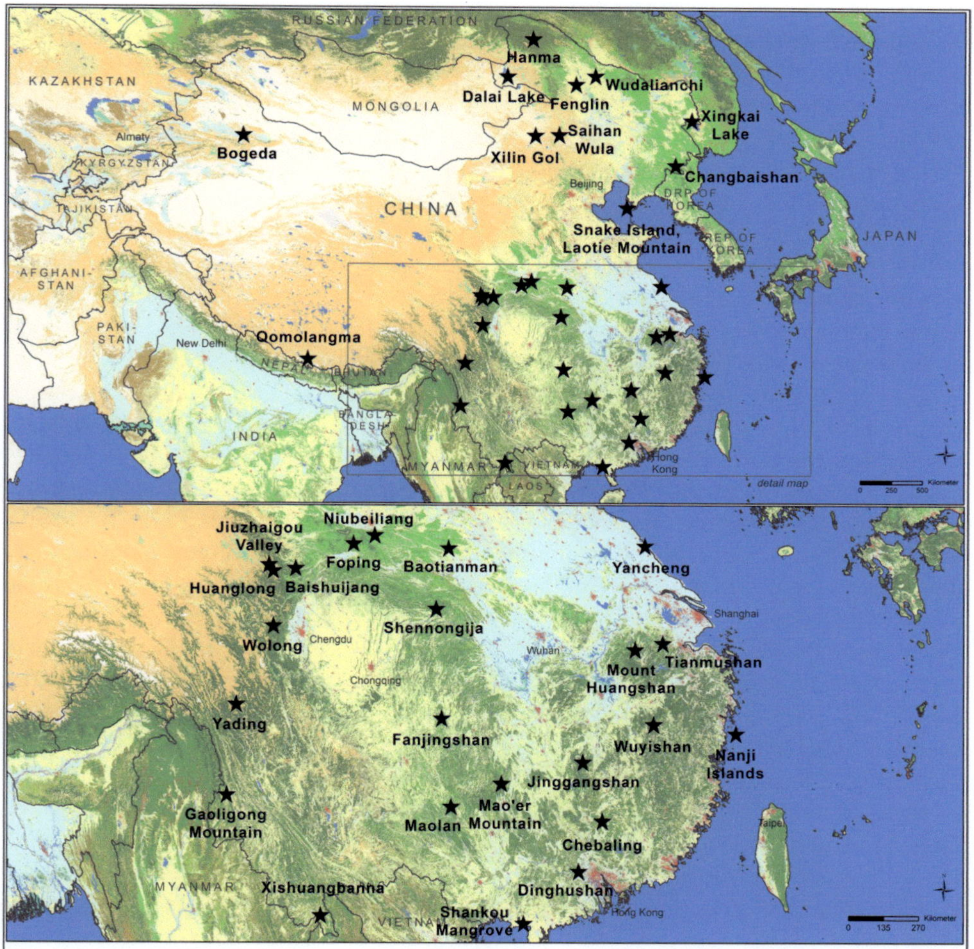

★ Chinese UNESCO Biosphere Reserves — National boundaries ▭ Land cover 2015 (for classification see annex 1)

	Designation	*Periodic Reviews*	
Changbaishan	1979	1998	2017
Dinghushan	1979	1998	2017
Wolong	1979	1998	2017
Fanjingshan	1986	1999	2017
Xilin Gol	1987	1999	2017
Wuyishan	1987	1999	2018
Bogeda	1990	2001	2016
Shennongija	1990	2001	2017
Yancheng	1992	2003	2015
Xishuangbanna	1993	2004	2016
Maolan	1996	2007	
Tianmushan	1996	2007	
Fenglin	1997	2009	
Jiuzhaigou Valley	1997	2011	
Nanji Islands	1998	2011	
Shankou Mangrove	2000	2011	
Baishuijang	2000	2011	
Gaoligong Mountain	2000	2011	
Huanglong	2000	2011	
Baotianman	2001	2012	

	Designation	*Periodic Reviews*
Saihan Wula	2001	2012
Dalai Lake	2002	2014
Wudalianchi	2003	2014
Yading	2003	2014
Foping	2004	2018
Qomolangma	2004	2019
Chebaling	2007	
Xingkai Lake	2007	2019
Mao'er Mountain	2011	
Jinggangshan	2012	
Niubeiliang	2012	
Snake Island, Laotie Mountain	2013	
Hanma	2015	
Mount Huangshan	2018	

Source: UNEP-WCMC, 2020b & UNESCO, 2020c
Layout: M. Engelbauer, December 2020

The model of biosphere reserves for nature conservation in China has attracted wider national interest, which led to China MAB's decision in 1993 to establish a national network of Chinese biosphere reserves along and beyond the sites designated by UNESCO (Ding & Ounli, 2020, p. 217). Over time, the national network has grown to 177 national biosphere reserves, which serve as a platform to connect different nature reserves, share ideas and standards from the UNESCO biosphere reserve framework and participate in training courses. Hence, the process to nominate a site as a UNESCO biosphere reserve involves first participating in the national programme; then, there is a long selection process for candidates in the national network to be nominated for the World Network of Biosphere Reserves (19:19). Thanks to the Chinese network of biosphere reserves, a growing number of parties have shown interest in the MAB Programme and have participated in each step of the MAB Action Plan in hopes of meeting UNESCO standards in the future and have their site nominated for the World Network of Biosphere Reserves (19:4).

The periodic review process in China is quite sophisticated. The first step begins several months before the formal review. A small group of experts and staff from the China MAB Secretariat travel to the biosphere reserve and to assess the situation, provide feedback to managers on their performance and suggest improvements before the formal periodic review (19:6). During the formal periodic review, a larger group of experts and staff from the MAB China Secretariat travel to the biosphere reserve to perform a 10-year evaluation and attend meetings with local people and the staff of the protected area. Then, the experts hold an internal meeting to develop a joint review, which is subsequently presented to the biosphere reserve staff, leadership from the local government as well as the local villagers to discuss main achievements from the last 10 years and the direction that future activities should take. The summary of this process is then translated into English and submitted to UNESCO (19:5; 19:21). The comprehensive work undertaken for the periodic review process is fully funded by the central government (19:22).

Four Chinese biosphere reserves were listed in the Exit Strategy. One was the Dinghushan Biosphere Reserve, designated in 1979, which had extended issues with zonation and the involvement of local communities. At the time of the interviews in 2018, most of the problems had already been solved, and only one Chinese biosphere reserve remained affected by the Exit Strategy. A special task force was appointed to make the necessary adjustments and re-nominate the site (19:26).

Given China's size, a future goal is to add many more biosphere reserves to the UNESCO World Network to cover the entirety of the country's rich biodiversity and ecosystems (19:11; 19:20). China has a huge population and has faced high pressure to rapidly develop to meet its needs of the people. This development has heavily impacted the environment and ecosystems. Thus, the principle of co-existence between people and the environment within the MAB Programme is very welcome in China. This is also why the MAB National Committee has received significant support from the central government to promote biosphere reserves. Moreover, Chinese society has increasingly paid attention to wilderness and conservation issues. Over the next 10 years, there is a very high likelihood that China will continue to promote the MAB Programme and nominate more biosphere reserves (19:27; 19:30).

5.3.3 Indonesia

Figure 18: UNESCO Biosphere Reserves in Indonesia

★ Indonesian UNESCO Biosphere Reserves — National boundaries ☐ Land cover 2015 (for classification see annex 1)

	Designation	Periodic Reviews	
Cibodas	1977	2011	
Komodo	1977	1999	2013
Lore Lindu	1977	1999	2013
Tanjung Puting	1977	1998	2013
Gunung Leuser	1981	1999	2013
Siberut	1981	1999	2013
Giam Siak Kecil - Bukit Batu	2009		
Wakatobi	2012		
Bromo Tengger Semeru-Arjuno	2015		
Taka Bonerate-Kepulauan Seleyar	2015		
Belambangan	2016		
Berbak-Sembilang	2018		
Betung Kerihun Danau Sentarum Kapuas Hulu	2018		
Rinjani Lombok	2018		
Saleh-Moyo-Tambora	2019		
Togean Tojo Una-Una	2019		
Bunaken Tangkoko Minahasa	2020		
Karimunjawa-Jepara-Muria	2020		
Merapi Merbabu Menoreh	2020		

Kilometer 0 150 300

Source: UNEP-WCMC, 2020b & UNESCO, 2020c
Layout: M. Engelbauer, December 2020
© EuroGeographics for the administrative boundaries
© ESA Climate Change Initiative - Land Cover led by UCLouvain (2017)

Source: Own illustration

In Indonesia, the MAB Programme was initiated in 1972. The first four biosphere reserves – Cibodas, Komodo, Lore Lindu and Tanjun Putting – were designated in 1977. Four years later, two biosphere reserves were added: Gunung Leuser and Siberut (see Figure 18). Twenty-eight years later, in 2009, the Giam Siak Kecil-Bukit Batu Biosphere Reserve became the first site to be jointly initiated by the private sector and the local government, followed by the Wakatobi Biosphere Reserve in 2012 (SUDARMONOWATI, 2019, p. 2). From 2015 to 2020, 11 new biosphere reserves were nominated in a boom phase due to increased awareness amongst local governments and communities and a better outreach programme through improved communication about the concept of biosphere reserve (SUDARMONOWATI, 2019, p. 1).

The Indonesian MAB National Committee has a mandate to implement programme activities. The Indonesian government assigned the management of the MAB National Committee to the Indonesian Institute for Sciences. To perform its duties, the MAB National Committee coordinates with the UNESCO Regional Office in Jakarta and the Indonesian National Commission for UNESCO, which is linked to the Ministry of Education (PURWANTO et al., 2020, p. 44). So far, biosphere reserves are not included in national legislation; instead, all biosphere reserves are anchored in local or regional legislation by the governors of each district (10:5).

To fulfil all three functions of biosphere reserves and produce comprehensive programmes that safeguard natural resources for community welfare, the development of management plans as part of a multi-stakeholder approach is necessary (PURWANTO et al., 2020, p. 50). Management and action plans are elaborated for every five years for biosphere reserves in Indonesia. These are developed with all key stakeholders in the affected areas, such as national park or wildlife area authorities, forestry agencies, local governments, local communities, research institutes, universities and NGOs (10:6). However, coordination between these different stakeholders is one of the biggest challenges in the implementation of the MAB Programme. In some sites, only one administrative district is involved; in others, biosphere reserves span three different districts, which increases the complexity of stakeholder involvement and site management (10:4).

Another major challenge is securing sustainable funding implement the MAB Programme and related activities in biosphere reserves. Often, there are grants for three or four year-long projects to establish and strengthen local management institutions. However, once they end, it is a financial challenge for the management committee to continue operating (10:16).

In addition, changes in government and personnel for the management of the biosphere reserves is related to financial challenges. Local governments change nearly every five years. Some are very strong in terms of sustainability of processes, but others focus only on the economy, which challenges the continuity of the programme. This is not an issue for core areas, as they are managed by national park authorities; however, local governments have autonomy in buffer and transition zones. Moreover, the responsible minister and institutions within the MAB National Committee change up to every five years, which means coordination all these aspects are a significant challenge (10:17). Therefore, stakeholder coordination, staff turnover and a lack of sustainable funding are the most important constraints in the implementation of the MAB Programme in Indonesia (10:4).

With the Exit Strategy, several first-generation biosphere reserves had to be rezoned because they were identical with the core protected area, mostly a national park, lacking the buffer and transition zone (10:11).

To ensure the future of the biosphere reserves network in Indonesia, the governance system of the MAB Programme is more important than the sheer number of biosphere reserves. So far, there are several nominations in process, and the aim is to assess how many of Indonesia's 45 national parks could be developed as biosphere reserves (10:20).

5.3.4 Japan

Figure 19: UNESCO Biosphere Reserves in Japan

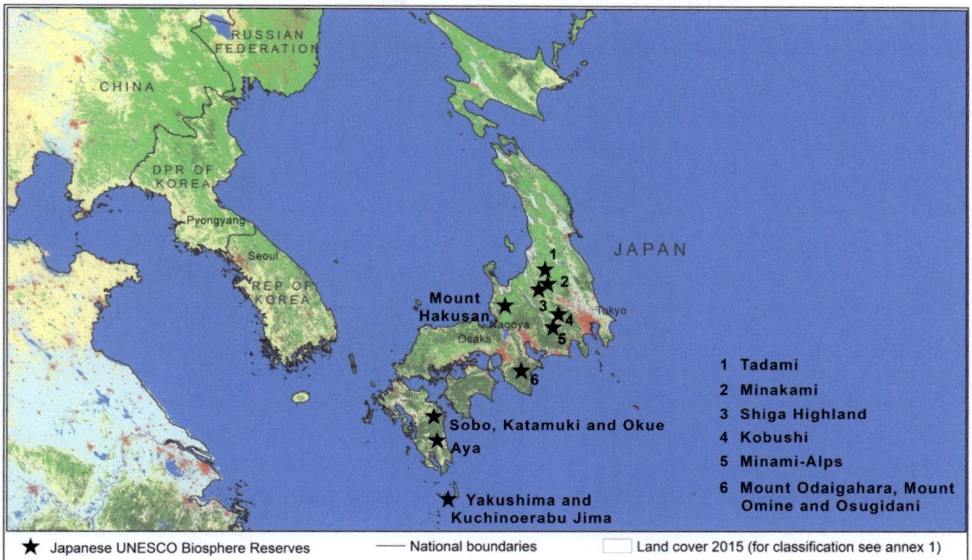

	Designation	Periodic Reviews	
Mount Hakusan	1980	1999	2016
Mount Odaigahara, Mount Omine and Osugidani	1980	1999	2016
Shiga Highland	1980	1999	2014
Yakushima and Kuchinoerabu Jima	1980	1999	2016
Aya	2012		
Minami-Alps	2014		
Tadami	2014		
Minakami	2017		
Sobo, Katamuki and Okue	2017		
Kobushi	2019		

Source: UNEP-WCMC, 2020b & UNESCO, 2020c
Layout: M. Engelbauer, December 2020
© EuroGeographics for the administrative boundaries
© ESA Climate Change Initiative - Land Cover led by UCLouvain (2017)

Source: Own illustration

In the late 1970s, the Japanese government asked the Environmental Agency to select candidate sites for nomination as UNESCO Biosphere Reserves. After consultations with the Forestry Agency, four sites were nominated and designated by UNESCO in 1980: Mount Hakusan Biosphere Reserve, Mount Odaigahara and Mount Omine Biosphere Reserve, Shiga Highland Biosphere Reserve and Yakushima Biosphere Reserve. Since the nomination process was almost exclusively conducted by the national government, biosphere reserves were not known by local communities, residents and the general public; thus, they did not play an integral role in nature conservation and sustainable development in Japan. All the sites contained core and buffer zones without any transition areas (MATSUDA et al., 2020, pp. 190-191).

After a long period of dormancy, the MAB Programme in Japan was gradually reactivated around 2010, when the Japanese National Commission for UNESCO decided to change the nomenclature for biosphere reserves in Japan to 'UNESCO Eco Parks'. At the same time, a collaborative initiative between the Forestry Agency, Miyazaki Prefecture, the town of Aya and two environmental NGOs examined the potential designation of a new biosphere reserve in Aya. The aim was to conserve and restore the endangered lucidophyllous forest and foster the sustainable development of local communities. Existing local activities precisely aligned with the concept of UNESCO Biosphere Reserves, and the site was successfully designated in 2012 (MATSUDA et al., 2020, pp. 192-193).

The designation process for Aya also forced the MAB National Committee to establish national criteria for the designation and evaluation of Japanese biosphere reserves in 2011. They stipulated that the management of biosphere reserves should be led by local governments, with broad participation from local stakeholders. Then, the Japanese MAB Secretariat urged local governments from the four existing sites to choose whether to retain the designation under the new conditions. All four biosphere reserves decided to maintain their nomination and established new management bodies called Biosphere Reserve Councils, which are led by local governments and involve a wide range of local stakeholders. In 2014, MABB ICC endorsed the extension of Shiga Highland Biosphere Reserve with a new transition area, and the other three sites followed in 2016 (MATSUDA et al., 2020, pp. 193-194). After Aya Biosphere Reserve, five biosphere reserves have been designated since 2014. Thus, there is currently a total of 10 sites in Japan (see Figure 19; 21:6).

In Japan, there are no dedicated laws for biosphere reserves, but there is the National Park Act and several other legislative systems for protected areas. These provide adequate protection for the core areas of the biosphere reserves. Since the concept of biosphere reserve combines different approaches, environmental and forestry agencies are particularly interested in them (21:7; 21:28).

The Japanese MAB National Committee is tied to the National Commission for UNESCO in Japan. The latter includes approximately 50 members chosen by the prime minister. Within this structure, there are seven sectors and different working groups. One working group in the natural science sector serves as the MAB National Committee in Japan. The Ministry of Education, Culture, Sports, Science and Technology serves as the MAB Secretariat in Japan (21:1). It mainly focuses on the selection process for new biosphere reserves in Japan. Meanwhile, site follow-ups and evaluations conducted by several voluntary research groups under the Japanese Coordinating Committee for MAB, a non-governmental and voluntary committee established by scientists. However, will need to change with the Lima Action Plan, which will be a challenging change in responsibilities for the MAB National Committee (21:3).

After the Madrid conference, the Exit Strategy and the quality improvement strategy provided a good opportunity to reactivate Japanese UNESCO Biosphere Reserves. Previously, the sites had been paper parks with almost no actions or activities. After the first four biosphere reserves were selected in a top-down approach, the inclusion of municipalities and local stakeholders in all Japanese biosphere reserves was seen as a positive change (21:5). The involvement of local stakeholders was also important for the periodic review process. However, the 10-year cycle for the periodic reviews was too infrequent. The Japanese MAB National Committee has established the idea to perform reviews every year and they requested to all biosphere reserves to give a short progress report (21:10).

In addition, every year a selection process is started for aspiring biosphere reserves to submit application documents to the MAB National Committee. The nomination process can take several years, as it now follows a bottom-up approach (21:28). The aim is to make biosphere reserves in Japan into sites for sustainable development by using local natural resources. Biosphere reserves offer a very interesting potential benefit for municipalities and local communities: ecotourism and ecolabelling (21:27). Thus, the Japanese biosphere reserves may also become important for the CBD's national biodiversity strategy because they showcase how to live in harmony with nature (21:4).

5.3.5 Republic of Korea

Figure 20: UNESCO Biosphere Reserves in the Republic of Korea

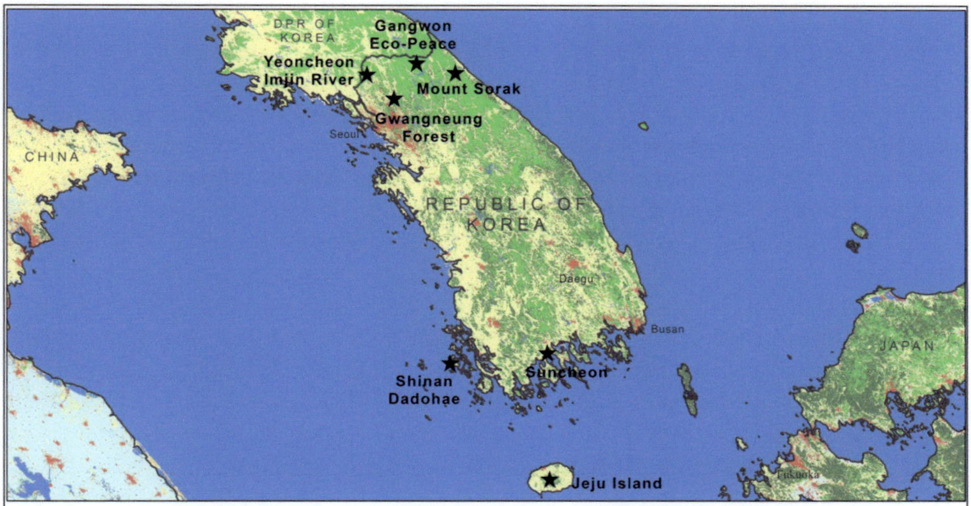

★ Korean UNESCO Biosphere Reserves — National boundaries ▭ Land cover 2015 (for classification see annex 1)

	Designation	Periodic Reviews	
Mount Sorak	1982	1998	2013
Jeju Island	2002	2013	
Shinan Dadohae	2009	2016	
Gwangneung Forest	2010		
Gochang	2013		
Suncheon	2018		
Gangwon Eco-Peace	2019		
Yeoncheon Imjin River	2019		

Kilometer 0 40 80

Source: UNEP-WCMC, 2020b & UNESCO, 2020c
Layout: M. Engelbauer, December 2020
© EuroGeographics for the administrative boundaries
© ESA Climate Change Initiative - Land Cover led by UCLouvain (2017)

Source: Own illustration

The MAB Programme in the Republic of Korea began in the early 1970s with the translation and publication of the UNESCO MAB Series. The MAB National Committee of the Republic of Korea was established in the 1980s and operated under the Korean National Commission for UNESCO. In 1982, Mount Sorak Biosphere Reserve was designated as the first Korean biosphere reserve (MAB National Committee of the Republic of Korea, 2015, p. 2). Like many other first-generation biosphere reserves, the biosphere reserve area was basically the boundary of a national park, and its main objective for the first 20 years was nature conservation and biodiversity protection (20:4). There were very few possibilities to implement sustainable development, as the site barely had a transition area (MAB NATIONAL COMMITTEE OF THE REPUBLIC OF KOREA, 2015, p. 1). Twenty years later, in 2002, the second Korean biosphere reserve, Jeju Island, was designated. From Jeju Island to the eighth biosphere reserve in 2019, designations for all new biosphere reserves have been initiated by local governments and supported by Korean society (20:4; see Figure 20). The MAB National Committee was relocated from the National UNESCO Commission to the Ministry of Environment in 2010, and the MAB Secretariat moved to the Korean National Park Service (MAB NATIONAL COMMITTEE OF THE REPUBLIC OF KOREA, 2015, p. 2).

One challenge for the MAB Programme and the biosphere reserves in the Republic of Korea is the existence of multiple administration sites, which involve different county governments with their own attitudes and vision. Some are eager to engage in the biosphere reserves, whilst others are not as actively involved (20:16).

The MAB Programme in the Republic of Korea also expanded domestic and international activities by supporting conservation within the Korean Demilitarized Zone and the establishment of regional and thematic networks. The Gangwon Eco-Peace and Yeoncheon Imjin River Biospere Reserves, which were newly nominated in 2019, are directly adjacent to the Korean Demilitarized Zone and seek to promote sustainable economic development through ecotourism and local product branding in areas that have long lagged behind in development due to the many legal regulations instituted after the Korean War in the 1950s. The Imhin River stands out as a notable example of conservation because of its topographical advantage; in addition, restrictions on civilian activities due to the heavy presence of military facilities in adjacent localities have left the river mostly untouched by humans, which has enabled an endemic Korean fish species to flourish. The MAB Secretariat of the Republic of Korea also supported the establishment of the regional East Asian Biosphere Reserve Network and the World Network of Island and Coastal Biosphere Reserves (MAB NATIONAL COMMITTEE OF THE REPUBLIC OF KOREA, 2015, p. 3; UNESCO, 2020c).

The MAB National Committee consists of 25 members, of which 15 are individuals and 10 are government officials or representatives of organizations such as UNESCO Korea, the Korean Ministry of Environment and the Ministry of Fisheries. All members are volunteers or perform their duties as part of their governmental work (20:5). Biosphere reserves have their own governance structure, with a management committee composed of local governmental institutions, local people, representatives of the MAB National Committee, local NGOs and initiatives and specialists from universities. Meetings are usually organized by local governments, which also finance the management of the sites. The national governmental is not paying for the salaries of biosphere reserve managers, but local governments can participate in governmental development programmes (20:10).

Biosphere reserves in the Republic of Korea are not legally recognized as protected areas. Only one sentence in the Environmental Conservation Act states that the central government must financially support biosphere reserves. There are currently discussions with the central government to change the environmental law to recognize biosphere reserves as domestic and nationally protected areas (20:6). The six South Korean biosphere reserves that were designated until 2018, have different protected area categories as their core area. Two sites have a national park as the core area, and the other four sites are provincial parks or provincial marine protected areas (20:7).

Since core areas are strictly protected areas, many people are concerned about their properties being restricted in future use of natural resources. One challenge derives from the name 'biosphere reserve' itself, which no longer reflects their current emphasis of the concept on sustainable development and conservation (20:2). The word 'reserve' implies strict protection; therefore, there are ongoing discussions about changing it, and different suggestions have been made for alternatives. Another difficulty is the translation of the term 'biosphere reserve' into Korean, which is quite long. In Korean, 'Jeju Island' is written with three characters, and 'biosphere reserve' is written with seven characters. Thus, together 'Jeju Island Biosphere Reserve' requires 10 characters (20:8).

Periodic reviews are prepared and organized by local governments with support from the MAB National Committee, as it includes different institutions and specialists in the social and natural sciences (20:11). The MAB National Committee of the Republic of Korea proposed that biosphere reserves submit a national review every year. Through this process, the biosphere reserve can quickly monitor conditions, recommendations made by the national committee and scientific data. Feedback from the biosphere reserves has been positive so far, as they must collect data and report to the central government as part of their normal duties (20:13).

The only first-generation biosphere reserve in the Republic of Korea, Mount Sorak, was affected by the Exit Strategy. To meet the requirements of the Statutory Framework, the biosphere reserve had to be expanded to include a larger transition area. The process was successful and the total area of the biosphere reserve nearly doubled (20:16).

With globalization, more local governments desire international cooperation and seek international designations such as World Heritage Sites, Biosphere Reserves and Global Geoparks. In Asia, the number of sites has increased in recent years; in the Republic of Korea, local governments are interested in designating new biosphere reserves (20:24).

5.4 Europe and North America

5.4.1 Austria

The Austrian MAB National Committee was established in 1973, only two years after the inception of the MAB Programme. It designated its first four biosphere reserves in 1977: Gossenköllesee Biosphere Reserve, Gurgler Kamm Biosphere Reserve, Untere Lobau Biosphere Reserve and Neusiedler See Biosphere Reserve. Mainly selected by scientists, these sites were predominantly areas for basic research, and did not focus on sustainable development. The first second-generation biosphere reserve, Großes Walsertal Biosphere Reserve, was established in 2000, followed by Wienerwald Biosphere Reserve in 2005 and Salzburger Lungau and Kärnter Nockberge Biosphere Reserves in 2012 (Köck & Arnberger, 2017, p. 87). Whilst all four first-generation biosphere reserves were withdrawn from the World Network of Biosphere Reserves during the Exit Strategy in 2014 and 2016, a new biosphere reserve was added in 2019: Unteres Murtal Biosphere Reserve. Thus, there are currently four UNESCO Biosphere Reserves in Austria (see Figure 21).

Figure 21: UNESCO Biosphere Reserves in Austria

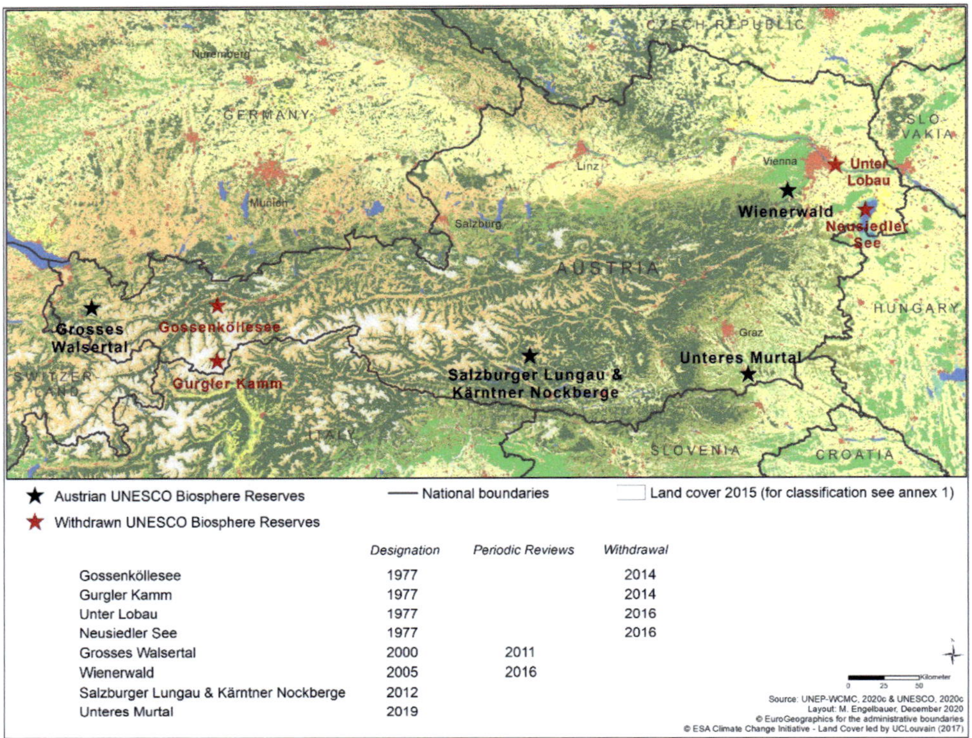

	Designation	*Periodic Reviews*	*Withdrawal*
Gossenköllesee	1977		2014
Gurgler Kamm	1977		2014
Unter Lobau	1977		2016
Neusiedler See	1977		2016
Grosses Walsertal	2000	2011	
Wienerwald	2005	2016	
Salzburger Lungau & Kärntner Nockberge	2012		
Unteres Murtal	2019		

Source: Own illustration

In Austria, biosphere reserves are called 'Biosphärenpark', which is German for 'biosphere park'. They are legally anchored in legislation of the individual federal states. So far, it has not been possible to integrate the biosphere reserves into federal law. National parks, however, are included under federal law; thus, national park administrators receive funding from the competent federal ministry. Therefore, it is highly unlikely that biosphere reserves will also be included under federal law because funds will probably not be increased; instead, national parks and biosphere reserves would have to share a budget. The integration of biosphere reserves' core zones under national legislation is relatively rigorous and complex in Austria. Core zones must be strictly protected nature reserves or areas and cannot be a Natura 2000 site or a natural monument (7:14). In addition, core zones must account for at least 5% of the total area (7:10). In Austria, the MAB Programme is coordinated by the Austrian Academy of Sciences, which is funded by the Federal Ministry of Education, Science and Research (Österreichisches Nationalkomitee MAB Programme, 2021).

Cooperation between biosphere reserves and the MAB National Committee is strongly encouraged in Austria, as representatives from all biosphere reserves also sit on the MAB National Committee. The latter meets twice per year, and one of the meetings is usually held at one of the biosphere reserves (7:7). Currently, the MAB National Committee is considering the creation of a joint brand for Austrian biosphere reserves. This means that they would have a common logo, appearance, publications and so on. However, there are still major hurdles, such as resistance from tourism companies that want to highlight their own regional or local logos and see no added value to have a joint brand. However, the federal states' political responsibilities make it difficult to agree on a common brand and presentation due to the various sources of political and financial support (7:8).

The Exit Strategy was an issue in Austria at a very early stage, as it had been clear for some time that there were two different quality levels in the country. The biosphere reserves of the first generation, were established in areas that had already been researched before. Consequently, they were just given another label for nature conservation without the designation of a buffer zone, without inhabitants and without a strategy. The second quality level consisted of modern, newly established biosphere reserves. Thus, it was difficult to substantially develop the country's seven biosphere reserves in a common and sustainable matter because they were so different. Subsequently, the Austrian National Committee developed its own strategy, and a research project was commissioned on how the four first-generation biosphere reserves could be transformed into post-Seville biosphere reserves. However, these transformations were not successful, as there was ultimately a lack of political will. For example, during the transformation of the Neusiedler See Biosphere Reserve, which only encompassed Lake Neusiedler itself, it became clear that the state government did not have the political will to implement the biosphere reserve concept. There was significant concern that the enlargement of the biosphere reserve would lead to more restrictions. Ultimately, the labels 'national park' and 'World Heritage Sites' were sufficient for the administration, and they did not mind withdrawing the site's designation as a biosphere reserve. In addition, the planned expansions were also not always well-received by local populations. For example, the Gurgler Kamm Biosphere Reserve has some of the best ski areas in the world, and it would have been difficult to coordinate its expansion in the presence of ski lifts. Thus, the sites' expansion and transformation into post-Seville biosphere reserves was not successful. As a result, the Gossenköllesee and Gurgler Kamm Biosphere Reserves were withdrawn from the World Network of Biosphere Reserves in 2014, and the Unter Lobau and Neusiedler See Biosphere Reserves followed in 2016 (7:20; 7:30).

In the global discussions about the Exit Strategy in the early 2010s, Austria introduced a proposal to designate biosphere reserves that had successfully engaged in research with a new label: MAB Research Sites. The idea was to give them their own label since they would no longer be part of the World Network of Biosphere Reserves but keep them under the umbrella of UNESCO. However, this proposal was rejected by the MAB ICC, and the Exit Strategy began (7:31).

In Austria, a key question for the MAB Programme and the World Network of Biosphere Reserves is how to distinguish biosphere reserves from well-known World Heritage Sites and national parks that have existed for over 130 years. Another challenge is the new label of 'geoparks', which closely resemble the concept of biosphere reserves (7:29).

Moreover, a major challenge for the management of biosphere reserves in Austria is the administration of sites across federal states. Due to their different responsibilities and the biosphere reserves' complete dependence on federal states for funding, the sites tend to be highly politicized (7:3). This should also be reflected in the composition of biosphere reserve administrations and the MAB National Committee. The managers and staff of the biosphere reserves should not act as mere executors for the ministries or federal states but rather try to involve local populations in their efforts. It is crucial to have a clear understanding of the functions of biosphere reserves beyond nature conservation, such as through the establishment of certified partner companies (7:27), the promotion of regional products and the use of a UNESCO biosphere reserve brand by certified accommodation providers and sustainable producers (7:29). However, it is also crucial for management to include interdisciplinary experts with knowledge about the law, the economy, spatial planning and biosphere reserves. This is one extremely important lesson learned that Austria had to make in the last years (7:28). Therefore, it is important to strengthen the management's position by establishing a clear framework before the designation of a site. In Austria, there are requirements in place for sufficient management and provision of staff and budget once a biosphere reserve is established. This way it is not possible that huge biosphere reserves are only managed by one person with almost no financial means (7:26).

5.4.2 Bulgaria

During the communist era, the state party of Bulgaria actively in joined international conventions and treaties. It was even written into the constitution that the international law would take precedence if national law contradicted it (16:3). Consequently, Bulgaria became a pioneering member of the MAB Programme in 1971 and designated 17 biosphere reserves in 1977 (see Figure 22). During that time, Bulgaria was one of the best represented countries in the World Network of Biosphere Reserves (16:5).

However, during the first decades, Bulgaria did not actively engage in the management of the first-generation biosphere reserves or adapt them to the paradigm shift after the Seville conference in 1996. Neither the state nor the responsible ministry directed any policy to the renovation and adjustment of biosphere reserves to the new criteria, zoning system and functions. In 2002, one of the biosphere reserves was withdrawn from the network, and the total number of Bulgarian biosphere reserves was reduced to 16. Fifteen out of 16 biosphere reserves were strict nature reserves and only fulfilled the conservation function, with some monitoring, education and scientific research. However, they did not include a transition area with a sustainable economic development function, which became increasingly important in the new concept of biosphere reserves (16:6).

One of the reasons for the neglect of the MAB Programme was the governments focus on the development of Natura 2000 sites. The Bulgarian state party had to develop and designate Natura 2000 sites as a condition for joining the EU in 2007. As a result, there was little time and commitment for meeting the MAB Programme's requirements. Because Natura 2000 was so time-sensitive, the relevant agencies and people were unprepared, many sites were designated over privately owned lands and some regulations were established for these areas. Although the process occurred quickly, it was not well coordinated. It is important to note that, after the end of the communist era in 1990, state-owned lands were returned to private owners; this is why all kind of land properties around protected areas now exist in Bulgaria. Regulations for the Natura 2000 sites were not as strict, but people were not happy about the additional restrictions on their newly obtained properties.

Currently, Bulgarians believe that additional designations will result in additional restrictions (16:7; 16:1; 15:11).

Figure 22: UNESCO Biosphere Reserves in Bulgaria

★ Bulgarian UNESCO Biosphere Reserves — National boundaries ☐ Land cover 2015 (for classification see annex 1)
★ Withdrawn UNESCO Biosphere Reserves

	Designation	Periodic Reviews	Withdrawal
Alibotouch	1977	2016	2020
Bistrichko Branichté	1977	2016	
Boitine	1977		*merged to Central Balkan BR in 2017
Djendema	1977		*merged to Central Balkan BR in 2017
Doupkata	1977		2017
Kamtchia	1977		2017
Koupena	1977		2017
Doupki-Djindjirtza	1977	2016	2020
Mantaritza	1977	2016	2020
Maritchini Ezera	1977		2002
Ouzounboudjak	1977	2017	
Parangalitza	1977	2016	2020
Srébarna	1977	2017	
Steneto	1977		*merged to Central Balkan BR in 2017
Tchervenata sténa	1977	2017	
Tchoupréné	1977	2016	
Tsaritchina	1977		*merged to Central Balkan BR in 2017
Central Balkan	2017		

Kilometer 0 25 50

Source: UNEP-WCMC, 2020c & UNESCO, 2020c
Layout: M. Engelbauer, December 2020
© EuroGeographics for the administrative boundaries
© ESA Climate Change Initiative - Land Cover led by UCLouvain (2017)

Source: Own illustration

In 2012, Bulgaria received the first letters from the MAB Secretariat about adjusting biosphere reserves to conform to the new requirements. After the adoption of the Exit Strategy one year later, the MAB Secretariat sent another letter to Bulgaria requesting the adjustment of all biosphere reserve to the new requirement; otherwise, Bulgaria would need to withdraw them from the World Network of Biosphere Reserves (16:8). Then, the Ministry of Environment and Water, the MAB National Committee and the Bulgarian Biodiversity Foundation worked together to renew the biosphere reserve network in Bulgaria. Although funding was provided by the ministry, there was only enough for some of Bulgaria's 16 biosphere reserves. In addition, many municipalities refused to participate in the renewal process, as they were strictly against having biosphere reserves. After many meetings with mayors, local communities and municipalities, four areas were selected for the renewal process: the Central Balkan, Srébarna, Tchervenata sténa and Ouzounboudjak Biosphere Reserves (16:9; 16:10).

Another challenge in the renewal of Bulgarian biosphere reserves was disagreement within the MAB National Committee. Two municipalities resisted changing the first-generation biosphere reserves into post-Seville biosphere reserves and submitted a letter of withdrawal from the World Network of Biosphere Reserves to the ministry. However, in Bulgaria, the system states that the MAB National Committee must first make a decision before a state government can submit a withdrawal to UNESCO. Thus, some members of the MAB National Committee ignored the wish of the municipalities and their request for a withdrawal. Consequently, first-generation biosphere reserves were still enlisted but could not become post-Seville biosphere reserve without willing participation from local communities; at the same time, they could not be withdrawn from the World Network of Biosphere Reserves because the MAB National Committee could not reach an agreement on the matter (16:11).

In 2016, the MAB National Committee nominated four new post-Seville biosphere reserves and withdrew the Doupkata, Kamtchia and Koupena Biosphere Reserves. Four first-generation biosphere reserves in the Central Balkan region were merged into a single large biosphere reserve. Five sites requested an extension of the deadline to decide upon a renewal process. A sixth biosphere reserve was a national park with two municipalities that did not want to become a post-Seville biosphere reserve (16:17). However, stakeholders at Bistrichko Branichté Biosphere Reserve near Sofia want to become a post-Seville site and have already started preparing the nomination form; it has a high likelihood of remaining in the World Network (16:18). After the expert interviews were conducted in 2018, four biosphere reserves were withdrawn from the World Network in 2020 and two more in 2021. Thus, there are currently four post-Seville biosphere reserves left in Bulgaria.

For Bulgaria, the adoption and implementation of the Exit Strategy was a positive initiative. For years, inclusion in the EU had the highest political priority and meant significant additional work, also in the field of nature conservation, and therefore little importance was given to biosphere reserves. The firm deadline of the Exit Strategy was a wake-up call for the Bulgarian government, and adaptation of the biosphere reserves became a new priority (16:27).

A few years ago, the Bulgarian MAB National Committee decided that the new post-Seville biosphere reserves would be called 'biosphere parks' because the word 'reserve' in Bulgaria referred to strict nature reserves. The latter are the most restricted category of protected areas in Bulgarian law and thus were perceived by people as new regimes and regulations in land use. To avoid raising concern over additional restrictions, the decision was made to call biosphere reserves 'parks'. Moreover, this decision drew a clear line between the old and new designations; thus, post-Seville sites would not be confused for first-generation sites, which focused only on the conservation function in very restricted protected areas (15:14; 15:15; 16:16).

The Central Balkan Biosphere Reserve offers an interesting case study of adaptation to the new concept of biosphere reserves. In the 1990s, nine strict nature reserves in IUCN

Category I and one national park in IUCN Category II were designated in the Central Balkan region. Four out of the nine strict nature reserves were also nominated as biosphere reserves in 1977 (15:5), which had serious problems after the adaptation of the Exit Strategy. After the rezonation of existing biosphere reserves, a new post-Seville biosphere reserve was established with the strict nature reserves as the core zone, the national park as the buffer zone and the areas around the national park, where communities lived, as the transition area (15:3). The mountains in the Central Balkan region divide Bulgaria into northern and southern parts, each with their own climate, vegetation and culture. Therefore, the biosphere reserve connects the north and the south and represents the entire region (15:6). Five out of nine bordering municipalities, the Forestry Department and the national park jointly submitted the site's nomination as a UNESCO Biosphere Reserve. In addition, there are plans to establish a dedicated NGO to manage the site, as biosphere reserves are not covered by any legal frameworks in Bulgaria (15:1). One of the first planned activities for the new biosphere reserve is to establish a trademark for the region in connection to the biosphere reserve. This is also expected to help the tourism industry, as UNESCO sites are world-famous. Moreover, some educational programmes for schools, visitors and locals around the park are planned (15:10).

5.4.3 France

There are 14 UNESCO Biosphere Reserves in France, two of which are overseas sites in Guadeloupe and French Polynesia (see Figure 23). Moreover, France contains two transboundary biosphere reserves: Northern Vosges-Pfälzerwald (with Germany) and Mont Viso (with Italy). Around two thirds of French biosphere reserves are terrestrial sites that mainly focus on forest ecosystems, agriculture and tourism, whilst the other third are entirely or partly located in coastal zones (MATHEVET & CIBIEN, 2020, p. 117). The first three biosphere reserves in France were designated in 1977 and include landscapes of high natural and scientific value, such as the Taiaro atoll and its lagoon in French Polynesia, the exceptional oak forest in the Fango Valley in Corsica and the Camargue biosphere reserve, which covers the Rhone delta. In the 1980s and 1990s, seven other biosphere reserves were designated; France mainly sought international recognition for their interdisciplinary research on the relationship between humans and nature. Between 2012 and 2015, four territorial projects were designated as biosphere reserves: the entire watershed of the Dordogne River (spanning over 24,000 $km^{2)}$, the intersection of coastal plains in Audomarois, the Flanders interior and the Artois hills, Mount Viso in the Alps and the typical Mediterranean landscape around the Gardon River gorge in the south of France (MATHEVET & CIBIEN, 2020, p. 117; UNESCO, 2020c).

Until 2016, biosphere reserves were not mentioned in any French law. In August 2016, a law for biodiversity regrowth, nature and landscapes was adopted, and biosphere reserves are considered specific sites for testing sustainable development in connection to the Seville Strategy and the Statutory Framework. Although no rules or obligations were written into the law, it recognised the international designation. Thus, a national or nature park, an

NGO or a group of municipalities can all express interest in the concept of biosphere reserves on an equal basis. This is a significant difference with the model of *parcs naturels régionaux* (regional nature parks) in France, which is similar in concept to biosphere reserves; however, it entails many constraints, rules and procedures established by law. Therefore, the new law intended to retain the flexibility of the biosphere reserve concept (17:16; 18:3). A practical advantage of enshrining biosphere reserves in law is that any party that undertake a large infrastructure project, such as the construction of a road or a dam, must now consider whether it is compatible with biosphere reserve (18:4).

Figure 23: UNESCO Biosphere Reserves in France

★ French UNESCO Biosphere Reserves — National boundaries Land cover 2015 (for classification see annex 1)

	Designation	Periodic Reviews		
Commune de Fakarava	1977	2006	2017	
Falasorma-Dui Sevi	1977	1998	2006	2017
Camargue (delta du Rhône)	1977	2000	2006	2017
Cévennes	1984	1998	2015	
Iles et Mer d'Iroise	1988	2001	2012	
Mont Ventoux	1990	2007	2018	
Archipel de la Guadeloupe	1992	2014		
Luberon-Lure	1997	2010		
Fontainebleau et du Gatinais	1998	2010		
Vosges du Nord Transboundary BR	1998	2000	2011	
Bassin de la Dordogne	2012			
Marais Audomarois	2013			
Mont-Viso Transboundary BR (France/Italy)	2013			
Gorges du Gardon	2015			

0 85 170 Kilometer

Source: UNEP-WCMC, 2020c & UNESCO, 2020c
Layout: M. Engelbauer, December 2020
© EuroGeographics for the administrative boundaries
© ESA Climate Change Initiative - Land Cover led by UCLouvain (2017)

Source: Own illustration

However, one challenge of the MAB Programme in France is to increase the visibility of biosphere reserves and make them more understandable to the public, ministries and national and local agencies. Biosphere reserves were associated with national parks until they were included in the law in 2016. However, biosphere reserves are more than a label; they are akin to a philosophy and provide a way of understanding sustainable development.

Many other protected area categories fulfil similar functions, but the idea of a cooperation between local management and scientific researchers originated from biosphere reserves. In France, there is a strong link between research and management, which makes biosphere reserves very special (18:2).

Since 2016, the MAB National Committee in France has existed as an association (NGO) called MAB France. Half of the members are from biosphere reserve organizations (permanent members), whilst the other half are from research institutions (co-members). The association also allows other members to participate in discussions, but they do not have voting rights. Biosphere reserves and research institutions build the management committee and choose members of the Secretariat, except for the chair. The chair is a scientist nominated by the Ministry of Foreign Affairs. This tradition was inherited from the previous system, in which the MAB National Committee was part of the French Commission for UNESCO (18:5; 18:8). The chair and vice-chairs have more a political representation of the association and are not paid for their work, as they are independent researchers. The director of MAB France addresses more technical issues and the management of everyday tasks, such as organizing meetings, preparing periodic reviews, developing projects and seeking funding. This position is funded by the French Agency for Biodiversity, which was also created in 2016 (18:9).

Most biosphere reserves in France are organized as *parcs naturels régionaux*, and some are hosted by the new national parks system, which has increasingly approached the idea of biosphere reserves through sustainable development and participatory processes. Each biosphere reserve is an association in which all main stakeholders are part of the organization. The biosphere reserve in French Polynesia has a distinct management system, but it is also a community-run site in which the mayor serves as the chair of the biosphere reserve (18:6).

The use of the term 'reserve' is the subject of considerable discussion in France. Some members of the MAB National Committee argue that it is associated with restrictions and loss of use rights, whilst others claim that it must be better explained in the future. At the national level, the name 'biosphere reserve' is used; however, at the local level, organizations and people are free to use other names. The most common one is simply 'biosphere', which also appears on their logo (18:7).

Periodic reviews are conducted by the staff of biosphere reserves with the support from the national committee and local students. In Toulouse, for example, there is a master's degree that focuses on the MAB Programme; therefore, a group of students often conduct discussions with locals during periodic reviews or even the nomination process for new sites (18:10). A particular challenge associated with the periodic reviews is the multitude of other evaluation processes in France for other protected area categories. For example, regional nature parks renew their statute every 15 years. New national parks also have a charter with a long and rigorous external evaluation process. Therefore, it is especially difficult if a biosphere reserve is also a regional nature park or national park; it may take three years to renew a charter, after which the periodic review process for biosphere reserves is due again (18:13).

In France, the Exit Strategy is considered a good quality improvement strategy for increasing the seriousness and organization of the biosphere reserve network. For French biosphere reserves, it was a crucial signal for some of the sites to not lose the label. The Exit

Strategy was a concrete pressure point that led them to react and take action. Two biosphere reserves in France did not fulfil the criteria of the Statutory Framework. The first one involved a misunderstanding about zonation; thus, a new application was sent to UNESCO after re-zonation. The second one involved the biosphere reserve in Corsica, which was considered too small and had no management plan or policy. In 2018, at the time of the expert interviews, the team from the biosphere reserve was in process of preparing a management policy because they did not want to lose the designation (18:14).

In many of the older biosphere reserves, local communities neither know that they live in a biosphere reserve nor understand what this means. This is a considerable problem and stems from low government support. The French government considers the biosphere reserves to be a scientific idea rather than a territorial planning model. However, this is now changing, as many in the French Agency for Biodiversity are convinced that biosphere reserves are a promising model for the future (18:18).

Increasing visibility of the biosphere reserves network in France is important for the designation of additional biosphere reserves in the future. With each new biosphere reserve, more politicians are involved and can then promote the concept and model in the government and parliament. This would be one way to increase awareness of and support for biosphere reserves. In recent years, the main focus of MAB France has been on improving the quality of biosphere reserves, and significant technical expertise has been gained. The current network requires strengthening, and new nominations would help this process (18:20).

5.4.4 Germany

In Germany, participation in the MAB Programme and the establishment of biosphere reserves were strongly linked to historical events, especially the country's division into East and West Germany and their different political systems. The Federal Republic of Germany and the former German Democratic Republic established separate MAB National Committees in 1972 and 1974. The establishment of large protected areas in the German Democratic Republic was not possible for a variety of reasons, but the first two biosphere reserves were recognised in 1979: the Steckby-Lödderitz forest (which is now part of the Flusslandschaft Elbe Biosphere Reserve) and the Vessertal forest (which is now the Thüringen Forest Biosphere Reserve). Their main objectives and functions were research and information and education services. By contrast, the favoured nature conservation category in the Federal Republic of Germany was national parks. Germany's first national park, Bayerischer Wald, which was established in the 1970s, was also nominated as a biosphere reserve in 1981. In both East and West Germany, biosphere reserves were of little public and political interest, and their potential to help people understand nature conservation and landscape development was not realised in the first decade (Nauber & Pokorny, 1994, p. 27; Mayerl, 2005, p. 23).

However, interest in biosphere reserves notably increased in 1990 in the German Democratic Republic when a national park programme was established shortly before its reunification with the Federal Republic of Germany. Both existing biosphere reserves were significantly expanded and, in addition, five national park, three nature parks and four new biosphere reserves were designated: Schorfheide-Chorin, Spree Forest, South-East Rügen and Rhön Biosphere Reserves. In Western Germany, two additional biosphere reserves were designated: the Wadden Sea and Berchtesgaden (Nauber & Pokorny, 1994, pp. 27-28; Mayerl, 2005, pp. 23-24). One very interesting site is the Rhön Biosphere Reserve, as it spans three federal states and the former East and West Germany, thus linking different nature conservation systems (Nauber & Pokorny, 1994, p. 33). In the 2000s, four more biosphere reserves were designated; they featured a clearer focus on the transition area and incorporated changes brought by the Seville Strategy and the Statutory Framework. Thus, there are currently 16 UNESCO Biosphere Reserves in Germany which represent a wide range of landscapes, from the Wadden Sea to the Alpine region (see Figure 24).

Figure 24: UNESCO Biosphere Reserves in Germany

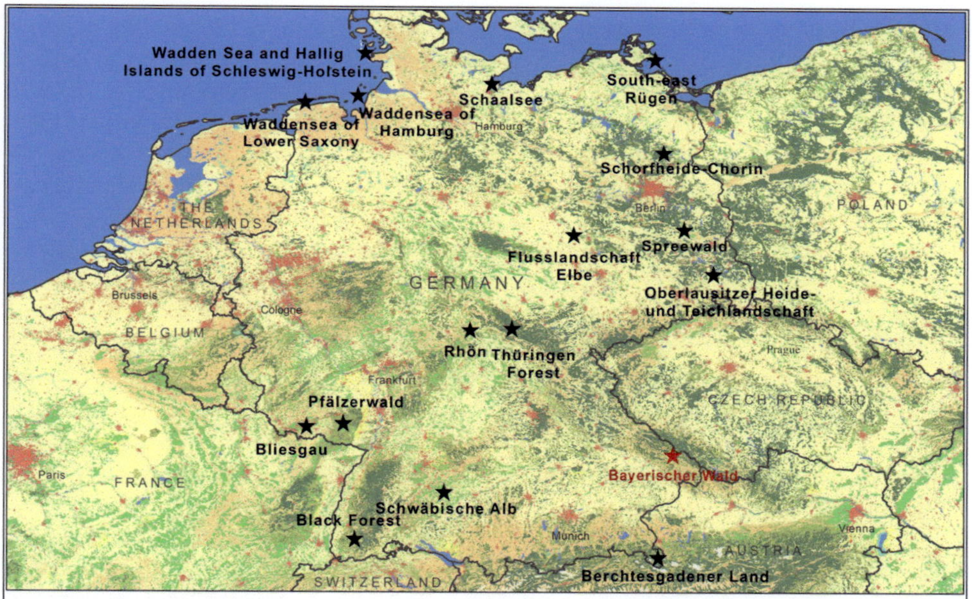

★ German UNESCO Biosphere Reserves — National boundaries ☐ Land cover 2015 (for classification see annex 1)

★ Withdrawn UNESCO Biosphere Reserves

	Designation	Periodic Reviews	Withdrawal
Flusslandschaft Elbe	1979	2002 2009 2018	
Thüringen Forest	1979	2002 2012	
Bayerischer Wald	1981		2007
Berchtesgadener Land	1990	2003 2013	
Wadden Sea and Hallig Islands of Schleswig-Holstein	1990	2006 2015	
Schorfheide-Chorin	1990	2003 2013	
Spreewald	1991	2003 2013	
South-east Rügen	1991	2004 2014	
Rhön	1991	2004 2014	
Waddensea of Lower Saxony	1992	2006 2015	
Waddensea of Hamburg	1992	2006 2015	
Oberlausitzer Heide- und Teichlandschaft	1996	2007 2017	
Schaalsee	2000	2012	
Vosges du Nord/Pfälzerwald Transboundary BR	1998	2004 2011 2014	
Bliesgau	2009		
Schwäbische Alb	2009		
Black Forest	2017		

Kilometer 0 40 80

Source: UNEP-WCMC, 2020c & UNESCO, 2020c
Layout: M. Engelbauer, December 2020
© EuroGeographics for the administrative boundaries
© ESA Climate Change Initiative - Land Cover led by UCLouvain (2017)

Source: Own illustration

A wide variety of actors have already recognised, applied and supported the concept of biosphere reserves in Germany. Thus, critical mass has already been reached, and biosphere reserves have become increasingly well-known (1:7). For example, there are two new initiatives to designate the Drömling and Rhine-Taunau areas near Wiesbaden as UNESCO Biosphere Reserve; stakeholders stumbled on the concept of biosphere reserves during an extensive search for a suitable model to manage the areas, which are not yet part of any nature conservation schemes. With regard to biosphere reserves, such bottom-up initiatives are becoming the new standard in Germany. More commonly, however, the managers of existing nature conservation areas seek a model to help preserve them for the future. For example, the newly designated Black Forest Biosphere Reserve originated from a large-scale nature conservation project. Thus, there are two main approaches to establishing biosphere reserves in Germany, each with their own justification (1:17).

In Germany, biosphere reserves are recognized in the Federal Act for the Protection of Nature; thus, they are part of the national protected area system (2:2). However, since Germany is a federal republic, the responsibility for designating biosphere reserves lies with individual federal states. Although biosphere reserves are defined differently in state legislation, they usually fall under the category of large-scale protected areas and refer to the Federal Act for the Protection of Nature (3:5, 2:8). One advantage of legally anchoring biosphere reserves in the protected area system is clear responsibility for their administration and management, which usually also entails financial and personnel resources. The allocation of departments is mainly assigned to the environmental ministries of federal states; therefore, specific parties are responsible for biosphere reserve (2:8). On the other hand, including biosphere reserves as a protected area category in the Federal Act for the Protection of Nature also creates difficulties with communicating the concept to members of the public, who are typically unfamiliar with it and may have had rather negative experiences with the designation of other protected areas, such as Natura 2000 sites, national parks and nature reserves. In such cases, they do not want another protected area designation, especially a 'reserve' (2:9).

Due to the inclusion of biosphere reserves in national legislation, it is possible that there are national biosphere reserves that are not recognised by UNESCO and thus not part of the World Network of Biosphere Reserves. Currently, UNESCO has only recognised 16 out of 18 German biosphere reserves. The Drömling Biosphere Reserve was designated under German state law in June 2019 and is currently preparing an application for designation as a UNESCO Biosphere Reserve. In addition, the Karstlandschaft Südharz Biosphere Reserve was designated under state law in 2009, but its designation as a UNESCO Biosphere Reserve remains pending due to opposition from a municipality within the biosphere reserve.

The MAB National Committee in Germany consists of 17 volunteer experts from science and practice and represent various thematic working fields. It is chaired by the Federal Ministry for the Environment, Nature Conservation and Nuclear Safety and managed by the German Federal Agency for Nature Conservation (BfN; BfN, 2020). Committee members of the committee are not paid for their work, which demonstrates the body's independence (2:31).

Over the last 20 years, it has been customary for the MAB National Committee to visit the biosphere reserves the year before the 10-year periodic reviews to write an independent evaluation report. In addition, it drafts an accompanying letter containing its recommendations to the MAB Secretariat for the periodic reviews (1:36). Thus, the MAB National Committee's evaluations and recommendations provide an external view on the work of German biosphere reserves. The MAB National Committee's ability to write recommendations and requests for individual state governments and biosphere reserves is a special feature of the German federal system. In addition, its evaluation reports are taken quite seriously (2:24). To date, discussions within the MAB National Committee remain strongly focused on core zones, as they must occupy least 3% of a biosphere reserve and legally fall under a strongly protected area category. For the future, the focus is on the transition zones to improve quality management in German biosphere reserves (2:25).

The transformation of the first-generation biosphere reserves was also an arduous process in Germany, as many of these areas (especially in West Germany) were previously national parks, which formed a double designation with biosphere reserves. One example was the Bayerischer Wald Biosphere Reserve, which was designated in 1981. The main reason for the designation was ecological research, as the core zone covered 75% of the biosphere reserve's total area and there was no transition area. In 1997, the national park was extended, but large parts of the local population opposed the enlargement due to concerns about an increase in bark beetles and their effects on forestry around the park. When the biosphere reserve was required to have a stronger focus on sustainable development and establish a transition area beyond the boundaries of the national park, there was also significant opposition from the local population. Because the model no longer fit the region, the Bayerischer Wald Biosphere Reserve was withdrawn from both the World Network of Biosphere Reserves and the German system of biosphere reserves in 2007 after discussions with local decision makers (WINKLER, 2019, pp. 174-175; 2:17; 2:48; 3:3).

A major challenge for the biosphere reserve network in Germany is the fact that many citizens believe that biosphere reserves are more or less the same as national parks and that the term is associated with prohibitions and restrictions; therefore, scepticism and reluctance from local people are high (1:20). Another challenge is the wide variation in financial and personnel resources at the administrations of biosphere reserves, in different federal states. Whilst some sites employ a high number of staff and even have specialists on their team (e.g. tourism development or education for sustainable development), other biosphere reserves are mainly staffed by a small number of conservationists and forestry experts. However, as part of having a model character; more testing is needed to determine how best to implement the concept of biosphere reserves within a national system (2:14).

Looking into the future, the question is how the model character of the biosphere reserves radiates into the surrounding regions to learn from each other in a global network but also to enable regions outside the biosphere reserve to learn from each other. Germany is a very diverse country, not only in terms of biodiversity, but also structural and cultural diversity. In this way, it is important to have more biosphere reserves (2:47). There are two main perspectives in discussions about whether Germany needs more biosphere reserves. On the one hand, the representativeness of natural and bioclimatic regions and ecosystem types is often mentioned. Some have suggested that the North German moors or the Alpine foothills should also be represented by biosphere reserves (JOB et al., 2019). On the other hand, discussions are less about the representativeness of natural areas and more about the relationships between humans and nature. The focus is more on landscapes with a certain charismatic natural land use, such as common pastures in the Black Forest, which are characterized by a strong sustainability approach, and are not yet covered by existing biosphere reserves in Germany (1:39).

5.4.5 Russian Federation

As in Germany, the creation and implementation of biosphere reserves in the Russian Federation is strongly linked to historical and political developments. In May 1972, the Soviet Union and the United States of America signed a cooperation agreement for environmental protection, and Soviet and U.S. American scientists began to conduct research projects on biosphere reserves (SOBISEVICH & SNYTKO, 2020, p. 3). In the Soviet Union, biosphere reserves were mainly established to integrate them in environmental monitoring as 'background stations' or ecological laboratories to determine how natural and semi-natural ecosystems respond to various levels of human intervention. Furthermore, the concept of biosphere reserves was used to promote international cooperation in the field of environmental monitoring, notably with the United States of America (SOKOLOV, 1981, p. 97). President Richard Nixon and Premier Leonid Brezhnev signed a bilateral agreement, in which both countries agreed to designate biosphere reserves in their countries as sites for major ecological research and environmental monitoring, as well as the preservation of natural ecosystems and gene pools. In 1976, the first Soviet-U.S. symposium on biosphere reserves was held in Moscow to exchange information and viewpoints on the implementation of the biosphere reserve concept (SOKOLOV, 1981, pp. 99-100).

In the Soviet Union, seven biosphere reserves were confirmed by the MAB ICC and included in the World Network of Biosphere Reserves in 1978. At the following General Conference of UNESCO in Paris, Soviet scientists proposed holding the first international conference on biosphere reserves. This became the First World Conference on Biosphere Reserves and took place in Minsk in 1983. Since the mid-1980s, there has been a tendency in Russia to transform nature reserves into biosphere reserves to represent all major biomes in the country (SOBISEVICH & SNYTKO, 2020, pp. 3-4). During the Soviet Union era, there were numerous monitoring stations across the country, from Europe to Japan. However, these did not have their own budgets and depended on the state and ministers. Since the collapse of the Soviet Union, many monitoring stations have been closed and much has changed, especially in terms of funding and cooperation opportunities (33:8).

In Russia, there have been attempts to develop a special law for biosphere reserves to clarify the structure and responsibilities. However, the only relevant legislation to date is the Law on Specially Protected Natural Areas (1995), which contains only two sentences about biosphere reserves (33:7). First, biosphere reserves are mentioned in the first category of protected land areas, *zapovedniks* (strict nature reserves), which are classified as IUCN Category I sites (OSTERGREN, 2001, p. 133). Second, it notes that 21 *zapovedniks* were part of the UNESCO Biosphere Reserves Programme in 1995 (OSTERGREN, 2001, p. 136). However, it does not mention support for participation in the international programme or how to organize biosphere reserves in the Russian Federation (33:7).

Figure 25: UNESCO Biosphere Reserves in Russian Federation

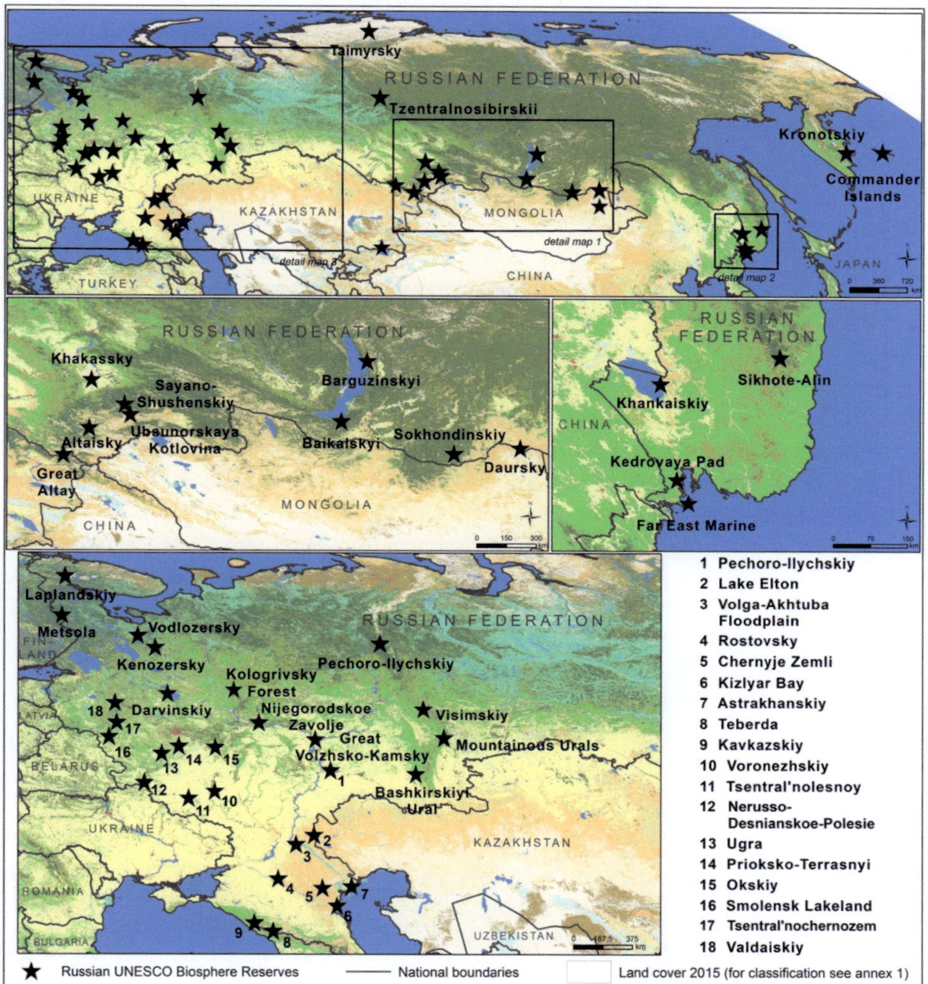

	Designation	Periodic Reviews	
Kavkazskiy	1978	1998	
Okskiy	1978	1998	
Sikhote-Alin	1978	1998	
Prioksko-Terrasnyi	1978	1999	
Tsentral'nochernozem	1978	1999	2016
Astrakhanskiy	1984	2005	2017
Kronotskiy	1984	2005	
Laplandskiy	1984	1999	2016
Pechoro-Ilychskiy	1984	1999	2017
Sayano-Shushenskiy	1984	1999	2017
Sokhondinskiy	1984	1998	
Voronezhskiy	1984	1998	2016
Tsentral'nolesnoy	1985	1999	
Baikalskyi	1986	2005	
Tzentralnosibirskii	1986	2005	
Chernyje Zemli	1993	2009	2018
Taimyrsky	1995	2009	
Ubsunorskaya Kotlovina	1997	2009	
Daursky	1997	2009	
Teberda	1997	2009	
Barguzinskyi	1986	2012	
Nerusso-Desnianskoe-Polesie	2001	2014	
Visimskiy	2001	2014	
Vodlozersky	2001	2012	
Commander Islands	2002	2014	
Darvinskiy	2002	2013	
Nijegorodskoe Zavolje	2002	2014	
Smolensk Lakeland	2002	2013	2018
Ugra	2002	2013	
Far East Marine	2003	2014	
Kedrovaya Pad	2004	2016	
Kenozersky	2004	2017	
Valdaiskiy	2004	2017	
Khankaiskiy	2005	2017	
Middle-Volga Integrated	2005	2017	
Great Volzhsko-Kamsky	2005	2017	
Rostovsky	2008	2017	
Altaisky	2009		
Volga-Akhtuba Floodplain	2001		
Bashkirskiyi Ural	2012		
Great Altay Transboundary BR	2017		
Khakassky	2017		
Kizlyar Bay	2017		
Metsola	2017		
Mountainous Urals	2018		
Lake Elton	2019		
Kologrivsky Forest	2020		

Sources: UNEP-WCMC, 2020c & UNESCO, 2020c
Layout: M. Engelbauer, December 2020

Zapovedniks are unique to the Russian Federation and the former Soviet Union. They represent the primary type of nature conservation area and function as scientific research sites. Most of them have very restricted public access and consist of undisturbed landscapes for scientific research (PRYDE, 1997, pp. 63-64).

The Russian Federation currently has 47 UNESCO Biosphere Reserves (see Figure 25). Given the country's large size, biosphere reserves are classified into different MAB regional sub-networks. For example, 28 out of 47 biosphere reserves belong to EuroMAB and participate in its regional meetings if funding is available and the meetings are within reach. Eastern biosphere reserves, on the other hand, cooperate with Asian countries in the East Asian Biosphere Reserve Network, which was established by the Republic of Korea and China and subsequently joined by Japan, the Democratic Republic of Korea, Mongolia and Kazakhstan. Finally, 18 Russian biosphere reserves are located around mountains and actively participate in the thematic sub-network of mountain biosphere reserves (33:3).

The Russian Federation supports the idea of the network of UNESCO chairs and try to link them to biosphere reserves. There are currently 64 UNESCO chairs in Russia, many of which are related to language and history and some to the natural sciences. New biosphere reserves should engage in cooperation with UNESCO chairs and youth. If young people can be reached, some might be good candidates for staff positions to work on biosphere reserves. Many young people from rural settlements are increasingly attached to nature and want to protect it. In addition, each biosphere reserve has a website; some contain rich information, which can significantly support the development of tourism (33:6).

A major challenge for the Russian biosphere reserve network is frequent changes in governments, ministers and heads of environmental protection agencies. With each new administration, cooperation must begin anew, and it is challenging to transfer knowledge about UNESCO Biosphere Reserves to the next government (33:15).

5.4.6 Slovakia

In the mid-1970s, Czechoslovakia founded a MAB National Committee and declared the first biosphere reserve in 1977: the Slovensky Kras (Slovak Karst) Biosphere Reserve (NOLTE, 2016, p. 62). At the time, most activities at biosphere reserves in Central Europe were dedicated to science, such as the production of biomass and the structure and ecological stability of forests. Because most biosphere reserves in this region are forests, the projects were mainly concerned with forestry practice and the utilization of results for science. The main job of the focal point for biosphere reserves during first-generation biosphere reserves was writing scientific articles and reports (5:18). This slowly changed in the 1990s, when humans focal to the programme; thus, main activities shifted to the use and management of natural resources, tourism development and the ways in which local populations could profit from activities (5:19). In 1990, the second biosphere reserve in Slovakia, Polana, was designated followed by two transboundary biosphere reserves, Tatra and Eastern Carpathian, in 1992 and 1998, respectively. Thus, there are currently four UNESCO Biosphere Reserves in Slovakia (see Figure 26).

Figure 26: UNESCO Biosphere Reserves in Slovakia

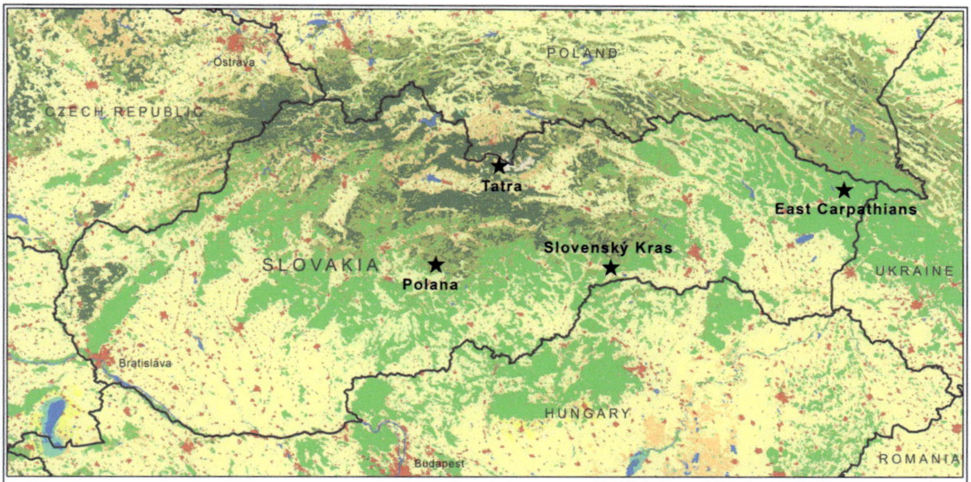

★ Slovak UNESCO Biosphere Reserves — National boundaries ☐ Land cover 2015 (for classification see annex 1)

	Designation	*Periodic Reviews*	
Slovenský Kras	1977	1998	2013
Polana	1990	2001	2013
Tatra Transboundary BR	1992	2014	2017
East Carpathians Transboundary BR	1998	2014	2017

Kilometer 0 15 30

Source: UNEP-WCMC, 2020c & UNESCO, 2020c
Layout: M. Engelbauer, December 2020
© EuroGeographics for the administrative boundaries
© ESA Climate Change Initiative - Land Cover led by UCLouvain (2017)

Source: Own illustration

The greatest challenge for the MAB Programme in Slovakia lies in the country's history and political changes. In the past, when Slovakia was still part of Czechoslovakia, the chairperson of the MAB National Committee was based in Prague and represented the entirety of Czechoslovakia. However, when the countries separated in 1993, a new government was formed in Slovakia, and a new Slovak MAB National Committee and new legislation for nature protection were established. One of the biggest challenges in the management of Slovak biosphere reserves is that they are not covered by legislation (5:2). Biosphere reserves in Slovakia are not specifically listed as protected areas under national nature protection law; they are only mentioned in the law as territories of international importance, and biosphere reserves are recognized as an international label in addition to national protection categories (NOLTE, 2016, p. 62; 5:4). The parties responsible for biosphere reserves are not directors of the biosphere reserves themselves but rather directors of national parks or of the protected landscape areas. Consequently, the management of biosphere reserves is not a priority for them (5:3). However, there are exceptions, such as the director of the Polana Biosphere Reserve, who is very committed to involving local farmers, state-owned forestry stakeholders and the owners of private forests in the management of the site. She won the Michel Batisse Award in 2017 (5:5).

In Slovakia, the political changes over the last 30 years have also had a significant impact on the management of biosphere reserves. In Czechoslovakia, all land and forest were owned by the state; they had been taken from private owners and municipalities and nationalised. Thus, the state taken over management of biosphere reserves. However, when Slovakia was established over 25 years ago, all the lands were returned to their previous owners. After living without any private property and valuables for years and then having their lands returned to them, many people built a house on the land for their families and cut down a quarter or even a third of the forest to sell wood. Some former landowners had already died, and their descendants lived in cities. Although they wanted their lands back, many did not want to work and live in remote areas. Consequently, many owners of lands located in protected areas and biosphere reserves opposed the protection status (5:8). As in the communist era, Slovakians were not consulted about whether they wanted be part of the biosphere reserves; after land titles were returned to private owners, there were major challenges related to the involvement of local populations (5:12). However, the Polana Biosphere Reserves once again provides a positive example. Locals in the area kept their traditions and religion during the communist era and did not give up their agricultural lands to cooperative farms. Thus, they were happy to be part of the biosphere reserve and saw it as beneficial to their lives (5:11).

Until 2016, the MAB National Committee in Slovakia mainly consisted of scientists from the Slovakian Academy of Science and universities. However, to adapt to the MAB Programme's new and more practical approach and address topics such as tourism, the involvement of local populations and the administration of villages within the biosphere reserves, scientists were increasingly replaced by practitioners from relevant fields (5:20). In addition, the director of the State Nature Conservation Agency became a member of the Slovak MAB National Committee, which is crucial to ensure the importance of biosphere reserves within the national protected area system. Moreover, the minister of foreign affairs was the new chair of the Slovak Commission for UNESCO. This support from high-level politicians was important for nature conservation and the management of internationally listed Slovak biosphere reserves (5:6).

Slovak biosphere reserves were also affected by the Exit Strategy. Amongst other criticisms, they were accused of not involving local populations. This was very problematic for biosphere reserves in Slovakia because of their lack of legal standing and political power. In addition, there was no funding to pay people to assist in necessary activities for the management of biosphere reserves according to the post-Seville criteria. In preparation for the General Conference, the ambassador for Slovakia brought a letter from the minister of the environment, which asserted that the situation of biosphere reserves in Slovakia would be improved, and they would be integrated into legislation. There was a great appreciation during the meeting. Although the letter attracted praise, the situation changed afterwards, and another four ministers came and went. At least the words of the minister were kept in mind in preparation for a new law about nature protection, where it is written that internationally protected areas such as World Heritage Sites, Natura 2000 areas and biosphere reserves should be considered with special care (5:4).

5.4.7 Spain

Spain has been involved in the development of the MAB Programme from the very beginning and designated its first biosphere reserves as early as 1977. After the World Conference on Environment and Development in Rio de Janeiro in 1992, Spain, and the MAB

Secretariat reflected on the implementation of Agenda 21 for the MAB Programme; Spain offered to host the second World Congress of Biosphere Reserves in Seville in 1995. Thirteen years later, Spain invited the member states of the MAB Programme to a third World Congress in Madrid. This shows Spain's clear great commitment and interest in the concept of UNESCO Biosphere Reserves (Clüsener-Godt, 2012, p. 437; Gómez Sal, 2018, p. 68). Since 1977, the number of biosphere reserves in Spain considerably increased; in 2020, 52 UNESCO Biosphere Reserves were designated in Spain, including three transboundary sites and even one transcontinental site with Morocco. As a result, Spain is the country with the most biosphere reserves in the world (see Figure 27).

Spain's 52 biosphere reserves can be divided into six groups. Ten sites are clustered around the theme of water and include both water-rich and water-scarce areas along the Atlantic coast; the Mediterranean Sea; the arid mountain biosphere reserve, Bardenas Reales, in the north; and the Mancha Húmeda in central Spain, which is one of the most valuable wetland ecosystems on the Iberian Peninsula. The second group is also strongly related to water, as it consists of islands: each has integral management over the entire island and receives high numbers of visitors. Seven out of eight island biosphere reserves are located in the Canary Islands: Lanzarote, Fuerteventura, Gran Canaria, Macizo de Anaga, La Gomera, Isla de Hierro and La Palma. The last one is Menorca on the Balearic Islands. The third group consists of 14 biosphere reserves in the intermediate mountainous regions of Spain. These areas are difficult to access and lower productivity than the Spanish plains, but much of their natural value is intact; thus, some mountain regions have also been designated as protected areas. The fourth category of high mountain sites consists of Ordesa-Viñamala Biosphere Reserve in the north of Spain and the Sierra Nevada Biosphere Reserve in the south. Their landscapes are dominated by cliffs and bare rocks; where agriculture is impossible, trees cannot grow, and plant species have adapted to these harsh climatic conditions. The fifth group consists of transboundary biosphere reserves; the Gêres-Xures, Meseta-Iberica and Tajo-Tejo Biosphere Reserves extend beyond the borders of Spain into mainland Portugal, and the Intercontinental Biosphere Reserve of the Mediterranean, designated in 2006, is shared with Morocco (RERB, 2020).

The sixth group consists of 12 biosphere reserves in the Cantabrian Mountains in the north and northeastern parts of Spain, which collectively form a nearly contiguous green carpet from east to west and host original forests and many emblematic species, such as capercaillie, wolves and bears. They comprise a mountain range that remains in a good state of conservation and runs from Galicia all the way to Cantabria and ends in the Picos de Europa. Some years ago, the idea was to create a large biosphere reserve in the greater Cantabrian area; all of the existing biosphere reserves would be merged into a single biosphere reserve. However, this did not happen due to the Spanish government's process of decentralisation in the 2000s. From this period onwards, the state was completely decentralised, and responsibilities were given to in the autonomous communities. In Grand Cantabria, there are now five autonomous communities: Galicia, Asturias, Cantabria, Castilla Leon and the Basque Country. Coordinating between all autonomous communities was and is very complex. The project of a single reserve has been lost in time, so it was thought to declare it little by little. Consequently, there are many small biosphere reserves in the area. All of them experience challenges related to viability and management because they are mountainous areas with a tremendous depopulation problem (35:4).

Biosphere reserves are recognised under Spanish legislation in the Law 42 on Natural Heritage and Biodiversity (2007). Three articles in this law are specific to biosphere reserves. Article 69 specifies the objectives of the Spanish network of biosphere reserves to maintain the interconnectivity of natural laboratories, to ensure the effective continuous comparison and transfer of information and to promote the generalisation of models of sustainable land-use planning and management (35:1; 35:2).

In Spain, the responsibility for the MAB Programme lies with the National Parks Autonomous Agency, which is part of the Ministry for the Ecological Transition and the Demographic Challenge. In addition, Royal Decree 387 of 2013 regulates the 34-member Spanish MAB National Committee. Its functions include institutional coordination, the application and approval of new sites and the monitoring of existing biosphere reserves. Additionally, it has two advisory bodies: a scientific council and a management council. The MAB National Committee includes staff from the state administration, such as the Ministry of Environment, the Ministry of Foreign Affairs, the Ministry of Rural Development, the Ministry of Agriculture, the Ministry of Research, the Ministry of Tourism; regional administration, including representatives of all autonomous communities with biosphere reserves, of which there are currently 16; and the local administration, such as local authorities, representatives of two NGOs and a representative of fishermen (35:3).

The management council consists of one technical representative from each biosphere reserve and is an advisory body to the MAB National Committee. The independent scientific council consists of only two representatives who prepare the periodic review reports. When a new biosphere reserve proposal is submitted, it is also sent to the scientific council. The council analyses the proposal and prepares a report. Then, the MAB National Committee sends the proposal and the scientific council's report to Paris to take the decision. In addition, the scientific council also monitors the biosphere reserves through an indicator system. The latter was approved in 2010 at a joint meeting of managers and scientists; system consists of eight indicators and is done annually. The results are sent to the MAB National Committee; in this way, it stays up to date on annual average value of the status of implementation of the MAB Programme for each site (35:5). In 2017, the analysis showed that 70% of Spanish biosphere reserves needed to improve their zoning and logistical function. Thus, the indicator system has been very useful and was made available to Portugal, Morocco and IberoMAB for any sites that want to use it (35:7).

The Las Cuentas Altas Manzanares Biosphere Reserve was the only Spanish biosphere reserve that was affected by the Exit Strategy due to zoning issues. In 2017, the site was rezoned and a new proposal was sent to the MAB IAC for evaluation (35:17).

In addition to the MAB National Committee and the two councils, the Spanish network registered a trademark for products with a characteristic logo. A special ruling in the royal decree regulates licencing for the use of such trademarks, which apply to products that originate from biosphere reserves. Quality monitoring for the products is conducted by the biosphere reserves themselves, whilst the dissemination of this brand image in newspapers and on television is overseen by the national MAB Secretariat and. In the future, there is an idea to also establish an electronic sales portal for products from biosphere reserves (35:8).

Figure 27: UNESCO Biosphere Reserves in Spain

1 Valles del Jubera, Leza, Cidavos y Alhama
2 Cuenca Alta del Río Manzanares
3 Real Sitio de San Ildefonso-El Espinar
4 Os Ancares Lucenses y Monte de Cervantes, Navia y Becerea
5 Muniellos
6 Los Valles de Omana y Luna

Source: UNEP-WCMC, 2020c & UNESCO, 2020c
Layout: M. Engelbauer, December 2020
© EuroGeographics for the administrative boundaries
© ESA Climate Change Initiative - Land Cover led by UCLouvain (2017)

★ Spanish UNESCO Biosphere Reserves —— National boundaries ▭ Land cover 2015 (for classification see annex 1)

	Designation	*Periodic Reviews*	
Sierra de Grazalema	1977	2004	2015
Ordesa-Viñamala	1977	2000	2011
Montseny	1978	2000	2011
Doñana	1980	2004	2012
Mancha Húmeda	1980	2002	
Sierras de Cazorla y Segura	1983	2004	2015
Marismas del Odiel	1983	2004	2017
La Palma	1983	2014	
Urdaibai	1984	2000	2011
Sierra Nevada	1986	2000	2012
Cuenca Altas del los Río Manzanares, ...	1992	2006	
Lanzarote	1993	2007	2018
Menorca	1993	2006	2017
Sierra de las Nieves	1995	2007	2017
Cabo de Gata-Nijar	1997	2011	
Isla de El Hierro	2000	2013	
Bardenas Reales	2000	2011	
Muniellos	2000	2014	
Somiedo	2000	2011	
Redes	2001	2013	
Dehesas de Sierra Morena	2002	2014	
Terras do Miño	2002	2014	
Valle de Laciana	2003	2014	
Picos de Europa	2003	2014	
Monfragüe	2003	2014	
Valles del Jubera, Leza, Cidavos y Alhama	2003	2014	

	Designation	*Periodic Reviews*
Babia	2004	2015
Área de Allariz	2005	2016
Gran Canaria	2005	2016
Sierra del Rincón	2005	2016
Los Valles de Omaña y Luna	2005	2016
Alto de Bernesga	2005	2016
Los Argüellos	2005	2016
Os Ancares Lucenses y Monte de Cervantes, ...	2006	2017
Los Ancares Leoneses	2006	2017
Las Sierras de Béjar y Francia	2006	2017
Intercontinental BR of the Mediterranean	2006	2017
Río Eo, Oscos y Terras de Buron	2007	2018
Fuerteventura	2009	
Geres/Xures Transboundary BR	2009	
La Gomera	2012	
Las Ubiñas-La Mesa	2012	
Marinas Coruñesas e Terras do Mandeo	2013	
Real Sitio de San Ildefonso-El Espinar	2013	
Terres de l'Ebre	2013	
Meseta Ibérica Transboundary BR	2015	
Macizo de Anaga	2015	
Tejo/Tajo Internacional Transboundary BR	2016	
Ponga	2018	
Alto Turia	2019	
La Siberia	2019	
Valle del Cabriel	2019	

Moreover, Spain has been involved in the establishment and management of biosphere reserves as part of its international cooperation efforts. It collaborates most closely with IberoMAB and EuroMAB regional networks. EuroMAB is a huge network that includes all European countries, the United States of America and Canada. Spain is only one of many. By contrast, Spain is much more active in IberoMAB and occupies a certain leadership position. Every year, the Spanish MAB network organises a seminar in collaboration with the Spanish Agency for International Cooperation. The location of the seminars rotates annually between Bolivia, Colombia, Guatemala and Uruguay (35:9).

One of the greatest challenges in Spain is confusion about the overlap between biosphere reserves and protected area such as national parks or nature parks. Thus, biosphere reserves do not have a clear identity and are not sufficiently distinguished from other protected area categories. Protected areas have a director, a management plan and a budget, but their objective is conservation. However, there is a lack of coordination between conservation authorities and those in charge of tourism, industry and rural development. Yet, this is precisely the spirit of biosphere reserves to combine all these activities in a joint action plan. Therefore, it is important to ensure that the biosphere reserves function as biosphere reserves, regardless of whether they are natural protected areas (35:12).

In Spain, it is widely understood that biosphere reserves are excellent models for spatial planning, as they allow regional stakeholders to decide where to engage in conservation, perform certain actions and allow the use of natural resources. However, there is no specific funding for this form of participatory management. External funding is always needed, such as EU rural development funds or EU agricultural funds (35:14). Moreover, the funding of biosphere reserve administration and management widely varies in Spain. There is not one single model. In Andalusia, Extremadura, Asturias and Madrid, autonomous authorities manage biosphere reserves; however, in these areas, the biosphere reserves coincide with protected areas (i.e. nature parks or national parks). Other sites are managed by local administrations in various ways. For example, in the Canary Islands, the island councils manage biosphere reserves as a local administration for all the municipalities of the island. By contrast, biosphere reserves are managed by provincial councils for each province in Catalonia, Barcelona and Galicia. Finally, some biosphere reserves are managed by a *mancomunidad.* A mancomunidad is an association of multiple municipalities joined to manage a biosphere reserve. However, one biosphere reserve is managed by an NGO: Castillo de Leon Biosphere Reserve in Leon in the province of Salamanca. The NGO is an association of cattle breeders that promoted the nomination and prepared the proposal; thus, the biosphere reserve was approved under its management (35:15).

5.4.8 Sweden

In Sweden, there are currently seven biosphere reserves recognized by UNESCO, all of which were designated after the Seville Strategy; six out of seven were designated in the last 10 years (see Figure 28; 27:4). They follow a strong bottom-up approach, with an adaptive co-management system that involves many stakeholders, including communities, enterprises and NGOs. This cooperative process is present from the initial stages to the establishment of the biosphere reserve, which usually takes more than five years until the nomination form is ready to send to UNESCO (32:3).

Figure 28: UNESCO Biosphere Reserves in Sweden

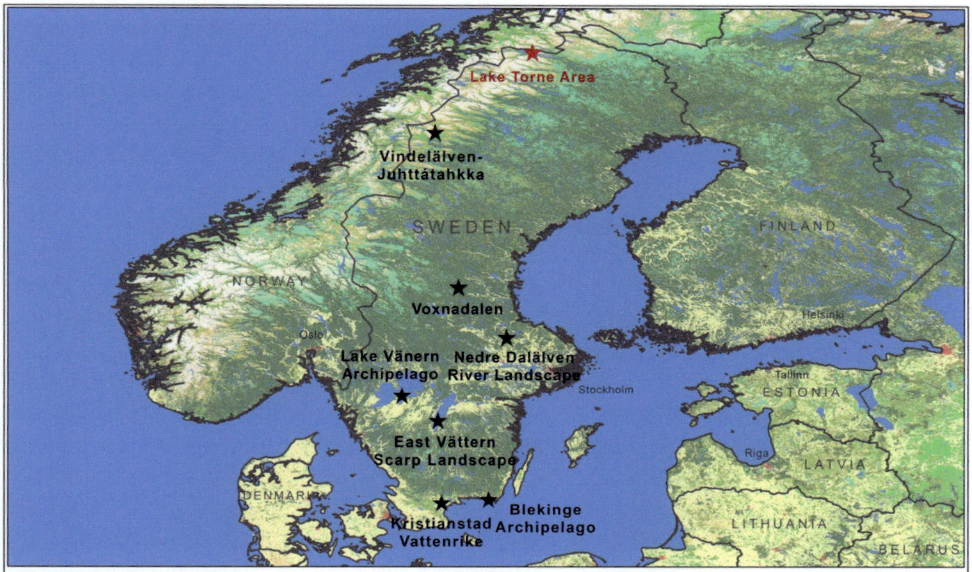

★ Swedish UNESCO Biosphere Reserves
★ Withdrawn UNESCO Biosphere Reserves
— National boundaries
☐ Land cover 2015 (for classification see annex 1)

	Designation	*Periodic Reviews*	*Withdrawal*
Lake Torne Area	1986	1999	2010
Kristianstad Vattenrike	2005	2016	
Lake Vänern Archipelago	2010		
Blekinge Archipelago	2011		
Nedre Dalälven River Landscape	2011		
East Vättern Scarp Landscape	2012		
Vindelälven-Juhttátahkka	2019		
Voxnadalen	2019		

0 87,5 175 km
Source: UNEP-WCMC, 2020c & UNESCO, 2020c
Layout: M. Engelbauer, December 2020
© EuroGeographics for the administrative boundaries
© ESA Climate Change Initiative - Land Cover led by UCLouvain (2017)

Source: Own illustration

The first biosphere reserve in Sweden was Lake Torne, situated in the northern most part of the country and was designated in 1986. However, there were never any specific biosphere reserve activities in the area after its nomination. Lake Torne Biosphere Reserve did not align with the criteria of the Seville Strategy. After more than 10 years of inactivity on the part of its administration, the decision was made to withdraw the biosphere reserve from the World Network of Biosphere Reserves in 2010 (27:5).

The current Swedish biosphere reserves represent different landscapes and socio-economic characteristics. Kristinstad Vattenrike Biosphere Reserve centres on a drainage basin of lakes and rivers that drain into the Baltic Sea. Economic activities in the area are related to agriculture, forestry and small- and medium-sized businesses. Lake Vänern Archipelago is located around the southern edge of Sweden's largest lake and is one of the most prosperous agricultural landscapes in the country. Blekinge Archigelago is in the south-eastern coastal area of the Baltic Sea; until recently, it mainly depended on the activities of the

Swedish Navy, with small-scale agriculture, fishery and forestry. Nedre Dalälven River Landscape, Voxnadalen and Vindelälven-Juhtatdahka are all river landscapes with wetlands and lakes. Finally, East Vättern Scarp Landscape has unique geological formations that create interesting ecozones. It lies in the central part of southern Sweden and is dominated by agriculture and forested lands, with villages and settlements consisting of small farms and individual homes (Kiellqvist et al., 2020, p. 104; UNESCO, 2020c).

The structure of biosphere reserves in Sweden varies by region. Several are organized as NGOs, which are responsible for the management of biosphere reserves in the area but not the territorial management of the geographic region. The biosphere reserve staff is also employed by the NGOs. One biosphere reserve is part of a municipal administration; it has an independent office funded by the municipality. Another site is a LEADER region. At least one person works full-time, with a maximum of eight staff members in one biosphere reserve (32:4). In addition, there is a national coordinator for the Swedish MAB Programme, who is employed by the Stockholm Resilience Center and contracted by the Environmental Protection Agency and the MAB National Committee (32:6).

In Sweden, perspectives on biosphere reserves differ from those in countries where biosphere reserves are a kind of protected area. Since Swedish legislation already specifies different types of protected areas, it was unnecessary to introduce another protected area category; instead, the focus was on connecting people with natural resources in both protected areas and non-protected areas, thereby achieving sustainable development (27:2). Core zones in biosphere reserves are always legally protected areas; since the establishment of the first post-Seville biosphere reserve in 2005, the focus was not on adding new restricted areas but rather working with existing initiatives around the area to increase understanding of why areas such as core zones need to be protected. Currently, there are intentions to include biosphere reserves in Swedish legislation. In fact, this is why biosphere reserves have been relatively successful so far; biosphere reserve managers do not represent nature protection agencies or impose new regulations for the region but rather engage in dialogue and collaborate with local stakeholders to increase understanding of how communities affect landscapes and can work in more sustainable ways (32:5; 27:6). In addition, the term 'biosphere reserve' is not used in Sweden, as the word 'reserve' is very negatively perceived. Therefore, biosphere reserves in Sweden are either called 'biosphere areas' or simply 'biospheres' (27:8; 32:7).

In 2010, the Swedish Environmental Protection Agency took over the responsibility as host for the MAB National Committee and has since funded the Swedish network of biosphere reserves and potential new sites (27:1). Previously, the MAB National Committee was a research council called the Swedish Academy for Natural Science (27:5). The MAB National Committee is relatively small, consisting of only five members from the Environmental Protection Agency, the Forest Agency and the Stockholm Resilience Center. The National Commission of UNESCO and the Agency for Ocean, Marine and Water also participate in meetings (32:9). There have been attempts to include more members in the committee (27:15). In addition to the MAB National Committee, there is also the biosphere reserve council, which consists of all coordinators of the biosphere reserves and the national coordinator. The role of the council is to communicate biosphere reserves' needs to the

MAB National Committee and serve as the advisory body to the MAB National Committee for new biosphere reserve nominations (32:10, 32:11).

So far, only one biosphere reserve in Sweden has completed a periodic review. The reporting interval of 10 years is useful for collaboration processes, as the Swedish approach requires long-term commitment and time to see any effects. For example, in one biosphere reserve, the local municipality decided to run all transport on locally produced renewable energy. The biosphere reserve let to this vision of the municipality and in these matters the perspective of 10 years for project implementation is a good reporting interval (32:14). However, periodic reviews also require substantial effort and entail many challenges with translating terminology from UNESCO documents and describing the Swedish context (27:16).

In addition to the 10-year periodic reviews, biosphere reserve coordinators hold telephone meetings with the national coordinator almost once a month and organise a yearly meeting with all coordinators. An annual meeting is also held within each biosphere reserve with all stakeholders from the municipality and different local organizations (27:10).

In 2018, the municipalities from one biosphere reserve organized their own conference for municipalities in biosphere reserves. It was clear that there was an enormous need for contact between different municipalities. For many people involved in managing biosphere reserves, it is good to have someone to talk to and not be alone with different or similar problems inside the biosphere reserves (27:12). The momentum associated with a growing network is a major opportunity for biosphere reserves to truly function as models for sustainable development (27:11).

There are other regions that would like to become UNESCO Biosphere Reserves. A preparation study was initiated for a region between Copenhagen and Malmö in Sweden and Denmark, respectively. In addition, two other initiatives have not yet been developed. Thus, it is possible that three or four new biosphere reserves will be designated in Sweden in the next 10 years (32:8).

5.4.9 United Kingdom

The United Kingdom was a member of the MAB ICC from 1971 to 1975, and many individuals were directly involved in the establishment and implementation of the MAB Programme in its early stages. One of the first 24 biosphere reserves designated by UNESCO in 1974 was Moor Mouse Upper Teesdale in the United Kingdom. Over the next three years, the United Kingdom nominated and designated another 12 sites. For a relatively small nation with a low diversity of biotic provinces (which were the site selection parameters at the time, according to UDVARDY), 13 biosphere reserves were a substantial number. In 1985, the United Kingdom withdrew from UNESCO, which weakened cooperation with the MAB Secretariat and hindered the work of the U.K. MAB National Committee. A study conducted in the same year as the Seville Conference in 1995 showed that agencies that were responsible for managing biosphere reserves had little interest in and commitment to the concept and that local populations had very limited awareness of biosphere reserves (PRICE, 2020, pp. 91-92).

Figure 29: UNESCO Biosphere Reserves in the United Kingdom

★ British UNESCO Biosphere Reserves — National boundaries ☐ Land cover 2015 (for classification see annex 1)

★ Withdrawn UNESCO Biosphere Reserves

	Designation	*Periodic Reviews*	*Withdrawal*
Wester Ross	1976	2016	former Beinn Eighe, extended and renamed in 2016
North Devon	1976	2016	extended and renamed in 2002
Biosffer Dyfi	1977	2009	
Loch Druidibeg	1977		2013
Moor House Upper Teesdale	1977		2012* formerly Cairnsmore of Fleet and Silver Flowe
North Norfolk Coast	1977		2014 Merrick Kells
Taynish	1977		2010
Galloway and southern Ayrshire	2012		
Brighton and Lewes Downs	2014		
Isle of Man	2016		
Isle of Wight	2019		

0 62.5 125 km

Source: UNEP-WCMC, 2020c & UNESCO, 2020c
Layout: M. Engelbauer, December 2020
© EuroGeographics for the administrative boundaries
© ESA Climate Change Initiative - Land Cover led by UCLouvain (2017)

Source: Own illustration

After the Seville conference, an analysis of the 13 biosphere reserves in the United Kingdom showed that barely any of the sites involved local stakeholders, none had a buffer zone or a transition zone and none had management plans. Although four biosphere reserves were important nature reserves and St. Kilda was a notable World Heritage Site, there were no resident populations in the area. As a result, four Scottish biosphere reserves – Caerlaverok, Claish Moss, Rum and St. Kilda – were voluntarily withdrawn from the World Network of Biosphere Reserves in 2002 (8:13; PRICE, 2020, p. 93) and two biosphere reserves – Braunton Burrows (renamed North Devon) in England and Dyfi in Wales – were accordingly extended (8:14; see Figure 29).

Ten years later, a second national review of biosphere reserves in the United Kingdom was conducted. Subsequently, four more biosphere reserves were withdrawn from the World Network: Taynish in Scotland in 2010, Moor House Upper Teesdale in 2012, Loch Druidibeg in 2013 and North Norfolk Coast in 2014. In addition, Beinn Eighe Biosphere

Reserve in Scotland was extended and renamed Wester Ross Biosphere Reserve in 2016, and the biosphere reserves formerly known as Cairnsmore of Fleet and Silver Flowe Merrick Kells were merged, extended and renamed Galloway and Southern Ayrshire Biosphere Reserve in 2012 (PRICE, 2020, p. 95; COETZER et al., 2014, pp. 89-90).

In the meantime, the managers of other sites considered the potential benefits of designation as a biosphere reserve. As a result, three new biosphere reserves were created through a bottom-up approach over the past five years despite a complete lack of funding from the national government (8:15). The first post-Seville biosphere reserve in the United Kingdom was Brighton and Lewes Downs, which was designated in 2014. Its origins were strongly politically motivated, as the Green Party was elected to and had a majority in the local parliament. It wanted to do something different and had the idea of establishing a biosphere reserve over the entire watershed, from a national park down the hills to the city of Brighton. Although the Green Party is no longer in power, the biosphere reserve still functions well and has a wide network of involved stakeholders. Therefore, it is a good example of a politically motivated biosphere reserve designation (8:31). Another new biosphere reserve was on the Isle of Man, in the middle of the Irish Sea. The Isle of Man is not legally part of the United Kingdom; rather, it is a self-governing Crown Dependency with its own government and parliament. However, the government supported the nomination of the biosphere reserve; as a result, all of the Isle of Man's land and sea area are designated as a biosphere reserve (PRICE, 2020, p. 97; 8:32). In 2019, the entirety of the Isle of Wight was also designated as a biosphere reserve. First and foremost, the local government and people saw this designation as a way to boost rural development. Whilst biosphere reserves were mainly selected based on nature conservation during the first years of the MAB Programme, this has completely changed; the focus of new nominations is now the regions themselves (8:33).

A major challenge for biosphere reserves in the United Kingdom is the requirement to include a much wider range of stakeholders and what this entails for power relations with parties such as the national park administration in protected core zones. Multiple actors are involved in the management of the biosphere reserve, and new governance structures are needed to reflect this. In addition, a major disadvantage of this development is the question who pays for the management of biosphere reserves. Thus, in the United Kingdom, there are a variety of governance systems (8:12).

As of 2018, the official government department that is responsible for biosphere reserves is the Department for Environment, Food and Rural Affairs, as it oversees all international designations and conventions, including Ramsar sites and the World Heritage Convention. However, this is only a small part of the work of one person in the department, and they do not usually have time to attend MAB National Committee meetings (8:4). In addition, biosphere reserves have no legal status under British law, and there has never been government funding for MAB activities (8:5). Although some human resources were available in the past, these have largely disappeared. All work and involvement with the MAB National Committee is performed on a voluntary basis (8:4). Biosphere reserves in Scotland are also not legally recognized; historically, they were anchored in the Department of the Environment, but it did not see much value in the concept. Around five years ago, the Department of Rural Development grew interested in biosphere reserves and started to invest in the re-designation of Beinn Eighe as the Wester Ross Biosphere Reserve. Subsequently, a project was granted by the EU and co-funded by the Department of Rural Development. Therefore, it is important to also involve the Department for Rural Development in biosphere reserve activities in England and Wales, as biosphere reserves are not only an environmental designation (8:10).

There are different governance structures for biosphere reserves in the United Kingdom. One approach is to create a foundation, an NGO or a charity that oversees the biosphere reserve and acts as an umbrella organization for different activities; it can be funded by different ministries or foundations. Other biosphere reserves are supported by local governments (8:11).

5.4.10 United States of America

In 1974, Nixon and Brezhnev signed a bilateral agreement on cooperation in environmental protection, specifically to support the MAB Programme, during a summit meeting between the United States of America and the Soviet Union in Moscow (13:3; BATISSE, 1982, p. 105). As a result of the joint agreement on the designation of biosphere reserves and a UNESCO MAB conference held in the United States of America, the United States of America selected 19 potential sites in 1974 and another nine sites in 1975 for designation as the first UNESCO Biosphere Reserves in the country. The selection was based on the sites' significance and representativeness of natural areas, as well as their long histories of biotic preservation and ecological research (FRANKLIN, 1977, pp. 264-265). UNESCO designated the first U.S. biosphere reserve in 1976, and all of them were managed by the National Park Service, the Forest Service or the Agricultural Research Service. The biosphere reserves mainly served as strictly protected areas for conservation and as benchmarks for monitoring ecological change and comparing the effects of human activities in the surrounding region. In the 1980s, the United States of America nominated regionally cooperative 'multisite' biosphere reserves, including two or more administrative units designated as a single biosphere reserve to foster regional cooperation and involve different agencies (U.S. MAN AND THE BIOSPHERE PROGRAM, 1995, p 4; GREGG, 1983, p. 26; U.S. MAN AND THE BIOSPHERE PROGRAM, 1994, p. 9). Between 1976 and 1991, the United States of America designated a total of 47 biosphere reserves (see Figure 30) – the most in the world at the time. Since 1991, however, the United States of America has not designated any more biosphere reserves.

In 1996, the U.S. MAB Programme came under fire due to a group called Sovereignty International Inc. The anti-UN organization claimed that biosphere reserves violated property rights and national sovereignty and that UNESCO Biosphere Reserves and World Heritage Sites were part of an UN experiment to take control of public and private lands in the United States of America. It strongly opposed the Convention on Biological Diversity and Agenda 21, convinced members of the U.S. Congress to cut funding for the U.S. MAB Programme and helped draft the American Land Sovereignty Protection Act, claiming that UNESCO prescribed policies and regulations that were not approved by elected officials in the United States or the U.S. Congress. During a hearing of the House of Representatives Committee on Resources in support of the American Land Sovereignty Protection Act, the group claimed that UNESCO, through the Statutory Framework and Article 4 to establish buffer zones and management plans regulating the use of the land around the core areas, was affecting land management decisions in the United States of America. The American Land Sovereignty Protection Act was passed by the House Committee with both Democratic and Republican votes, but it was not passed by the U.S. Senate (13:6; GILBERT, 2016, pp. 1-2).

Although the bill was not passed by the U.S. Congress and the Congressional Research Service found Sovereignty International Inc.'s allegations to be false, the process resulted in reduced political and financial support for the MAB Programme and the termination of many biosphere reserve activities. From its inception, the U.S. MAB Programme was co-led by the Forest Service and the National Park Service. This system functioned well for several years, until both agencies were pressured to withdraw their involvement in the MAB Programme by the U.S. Congress. Subsequently, the MAB Programme continued in the United States but without the support of these agencies (13:7; POFF, 1995, p. 63).

The Reagan administration had withdrawn the United States of America' membership and support from UNESCO in 1984. Nearly 20 years elapsed until President George W. Bush decided to rejoin UNESCO in 2003. The First Lady attended the General Conference in Paris and pledged support for all UNESCO programmes. After years of dormancy, the planning of the U.S. MAB Programme's renewal began. This was led by the Forest Service, which was also the chair of the U.S. MAB Programme. After the first meeting for the programme's renewal, Sovereignty International Inc. once again pushed the U.S. Congress to investigate the MAB Programme. Consequently, the Forest Service wrote an official letter to the U.S. National Commission for UNESCO and the State Department to notify them that it would withdraw from the MAB Programme. Thus, the U.S. MAB Programme once again lay fallow for many years (13:12).

Around 2013, under the presidency of Barack Obama, the State Department expressed interest in reactivating the MAB Programme to the director of the National Park Service. Since many biosphere reserves were under the jurisdiction of the National Park Service, it was appropriate for it to take the lead (12:1). To mark the return of the MAB Programme in the United States of America, a workshop was held at Rocky Mountain Biosphere Reserve in 2016 with former directors of the MAB Programme and biosphere reserve managers from Canada, Sweden and the United Kingdom. One of its main messages was to allow communities to be involved in the management of biosphere reserves; in addition, many discussions were held about periodic review processes (12:21). In recent years, the national network of biosphere reserves has been reinvigorated through numerous meetings and the attendance of several biosphere reserve representatives at EuroMAB meetings in Paris. However, under the administration of Donald Trump, tensions between the United States of America and UNESCO resurfaced, especially over the conflict in the Middle East. As of 1 January 2019, the United States of America has once again withdrawn from UNESCO. The United States will retain designations for its biosphere reserves, and some activities that would have taken place regardless of UNESCO membership due to the involvement of the National Park Service or regional economic development initiatives will continue. However, access to the international network will be restricted, and the United States of America longer use UNESCO branding or vote in the MAB ICC (12:7). In 2019, the MAB Secretariat confirmed that

Figure 30: UNESCO Biosphere Reserves in the United States of America

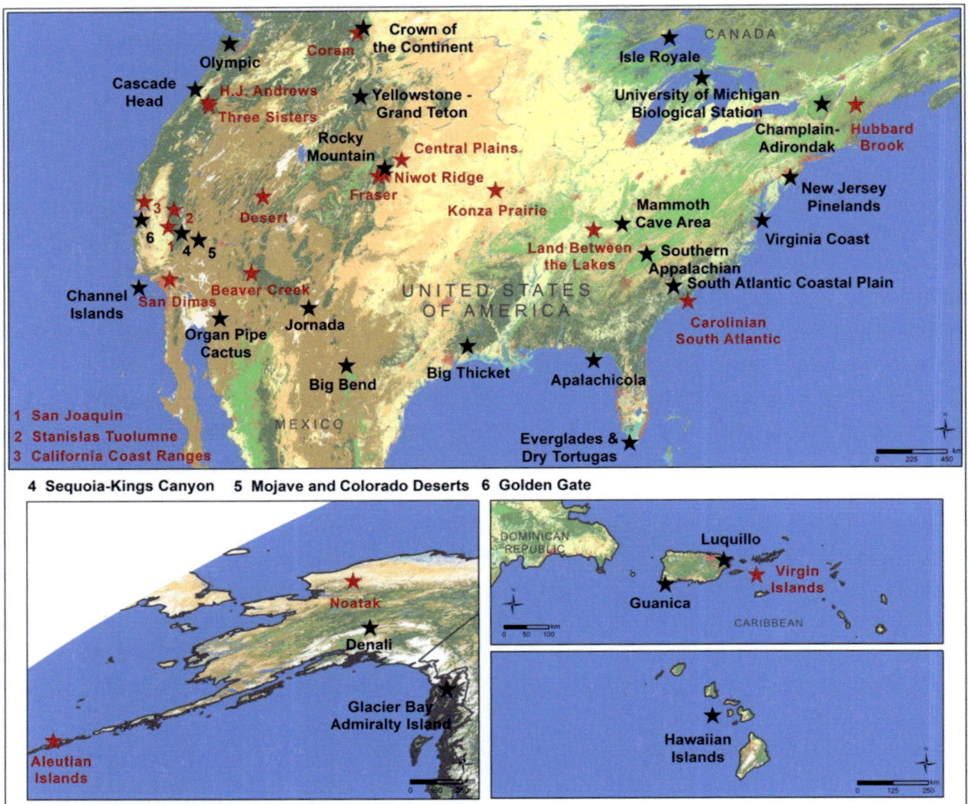

★ U.S. American UNESCO Biosphere Reserves
★ Withdrawn UNESCO Biosphere Reserves
Land cover 2015 (for classification see annex 1)

	Designation	*Periodic Review*	*Withdrawal*
Aleutian Islands	1976		2017
Beaver Creek	1976		2017
Big Bend	1976	2017	
Cascade Head	1976	2015	
Central Plains	1976		2017
Channel Islands	1976	2017	
Coram	1976		2017
Denali	1976	2017	
Desert	1976		2017
Everglades & Dry Tortugas	1976	2017	
Fraser	1976		2017
Crown of the Continent	1976	2014	
H.J. Andrews	1976		2017
Hubbard Brook	1976		2017
Jornada	1976	2017	
Luquillo	1976	2017	
Noatak	1976		2017
Olympic	1976	2017	
Organ Pipe Cactus	1976	2017	
Rocky Mountain	1976	2016	
San Dimas	1976		2018
San Joaquin	1976	2017	2019
Sequoia-Kings Canyon	1976	2017	
Stanislas Tuolumne	1976		2017
Three Sisters	1976		2017

	Designation	*Periodic Review*	*Withdrawal*
Virgin Islands	1976		2017
Yellowstone - Grand Teton	1976	2014	2018
Konza Prairie	1978		2017
Niwot Ridge	1979		2017
University of Michigan B. S.	1979	2014	
Virginia Coast	1979	2015	2018
Hawaiian Islands	1980	2017	
Isle Royale	1980	2017	
Big Thicket	1981	2017	
Guanica	1981	2018	
California Coast Ranges	1983		2017
Apalachicola	1983	2014	2018
South Atlantic Coastal Plain	1983	2014	
Mojave and Colorado Deserts	1984	2017	
Carolinian South Atlantic	1986		2017
Glacier Bay Admiralty Island	1986	2014	
Golden Gate	1988	2017	
New Jersey Pinelands	1988	2017	
Southern Appalachian	1988	2017	
Champlain-Adirondak	1986	2017	
Mammoth Cave Area	1990	2017	
Land Between the Lakes	1991		2017

Source: UNEP-WCMC, 2020f & UNESCO, 2020c
Layout: M. Engelbauer, December 2020
© EuroGeographics for the administrative boundaries
© ESA Climate Change Initiative - Land Cover led by UCLouvain (2017)

Source: Own illustration

> 'for the two countries that left the Organisation as a Member State (United States of America and Israel), biosphere reserves that already existed prior to their decision to leave, continue to exist in both countries, and monitoring will be maintained. They are even entitled to nominate new sites if they would want to do so, as this is in line with the objective to strengthen the MAB Programme as a means to achieve the SDGs'. (UNESCO, 2019, p. 24)

In the United States of America, the preparation of periodic reviews for biosphere reserves is one of the greatest challenges. Historically, there have been issues with zonation in the United States of America. The United States of America is sensitive about private property rights, especially the idea that international organizations could have any input on privately owned property. Moreover, the use of the word 'buffer' (as in 'buffer zone') is sensitive. During the periodic review process for Rocky Mountain Biosphere Reserve, its proposal to UNESCO included slightly different terms that would be more acceptable in the United States of America. The naming for core zones remained the same, but buffer zones were renamed 'areas of transition' or 'joint-management areas'. It was important to specify that buffer zones remained private property, but with a conservation easement that protects the core around the area. Moreover, transition zones were called 'areas of cooperation'; they were specifically named this way to reassure local populations that cooperative agreements would be strictly voluntary at the local, state and national levels and benefit the biosphere reserve as a whole (12:1). Rocky Mountain Biosphere Reserve used this new terminology in 2016; the following year, several other biosphere reserves in the United States of America followed its example and passed the periodic review process. However, 19 U.S. biosphere reserves stated that the current model of biosphere reserves is no longer what it was decades (i.e. mainly a science programme); as a result, they voluntarily withdrew their nominations from the World Network of Biosphere Reserves from 2017 to 2019. This was a necessary approach, as too many biosphere reserves had been inactive for too long and had no grass roots to further continue. However, the remaining biosphere reserves are in a different stage than the 19 voluntary withdrawn sites with local acceptance and local recognition (12:2). As of 2020, there were 28 biosphere reserves in the United States of America (UNESCO, 2020c, see Figure 30).

Another challenge in the United States of America is the branding of biosphere reserves. Many U.S. Americans know what a World Heritage Site is, but few have heard of biosphere reserves. It is crucial to clarify this, especially amongst local populations. There is a significant difference between in local and national perceptions of the UN and its different programmes and initiatives. The Trump administration was not as internationally focused, but at the local level activities depend on how liberal states are. On the East Coast or West Coast, there are few problems with recognition by the UN and the other thing is if people see an economic benefit in the designation. For example, the town of Paducah in Kentucky is designated as a UNESCO Creative City of Crafts and Folk Art, which has highlighted it as a destination for international travellers in a state that is typically wary of international organizations. However, a nomination was submitted for Paducah because they perceived economic benefits (12:5).

After some biosphere reserves were withdrawn from the World Network due to the Exit Strategy, a notable development was the formation of strong local groups that wanted to continue the work of biosphere reserves, especially in sites that were not managed by the National Park Service or the federal government. An example of good cooperation is the Champlain-Adirondak Biosphere Reserve in upstate New York, near the Canadian border. Joint meetings are held with Canadian biosphere reserves, and they are trying to implement joint local biosphere reserve activities (12:3). In addition, two biosphere reserves on the Mexican border – one in Arizona and one in Texas – are interested in working together as a potential transboundary biosphere reserve. The Mexican MAB National Committee is also interested in designating a transboundary site, but a change of administration is needed first (at time of interview in 2018; 12:12). A biosphere reserve in Hawaii is researching grant opportunities to promote the local practice of growing trees that are used in traditional canoes but were almost completely logged (12:20).

In the future, it will be important for the U.S. MAB Programme to further strengthen existing biosphere reserves and thus increase the quality of the entire network. This requires model sites that demonstrate to the world that the concept of biosphere reserves can function well and be useful. One example is the Mammoth Cave Biosphere Reserve, which was created around a common ground watershed; all participants agreed that it must be protected because many people depend on and consumes water from the site. Multiple counties, land users, representatives from the oil industry and others were brought together, and it really worked well. If there are enough success stories, biosphere reserves will be better understood (12:22). However, it is unlikely that there will be any nominations for new biosphere reserves in the United States of America over the next few years (12:23).

5.5 Latin America and the Caribbean

5.5.1 Colombia

In Colombia, there are five UNESCO Biosphere Reserves (see Figure 31). The first one is Cinturón Andino Biosphere Reserve in the Macizo Colombiano (Colombian Massif) in the Andes of south-central Colombia. It is the source of the Magdalena, Cauca, Patía, Caquetá and Putumayo Rivers and provides water to 70% of the country's population. It is also home to 13 ethnic groups. The biosphere reserve was designated in 1979, along with El Tuparro and Sierra Nevada de Santa Marta. El Tuparro Biosphere Reserve is in eastern Colombia near the border with Venezuela and is the only protected area in Colombia that represents the typical ecosystems of the Orinoquía, a transition zone between the highlands, the Amazon rainforest and the savannahs of the Guiana Shield. According to a study in *Science* (Le Saout et al., 2013), the Sierra Nevada de Santa Marta Biosphere Reserve in northern Colombia is one of the world's most valuable protected areas for the conservation of amphibian, bird, and mammal species. It stretches from the Caribbean coast, which encompasses preserved coral reefs and extensive beaches, to the Sierra Nevada, which reaches a height of 5,775 metres above sea level and is located only 42 km from the Caribbean coast. Its snowy peaks are considered sacred by the Arhuaco, Kogui and Wima indigenous people who live within the biosphere reserve (UNESCO, 2020c).

Figure 31: UNESCO Biosphere Reserves in Colombia

★ Colombian UNESCO Biosphere Reserves — National boundaries ☐ Land cover 2015 (for classification see annex 1)

	Designation	*Periodic Reviews*	
Cinturón Andino	1979	2001	2011
El Tuparro	1979	2001	2011
Sierra Nevada de Santa Marta	1979	2001	2011
Ciénaga Grande de Santa Marta	2000	2011	
Seaflower	2000	2011	

0 187,5 375 km

Source: UNEP-WCMC, 2020d & UNESCO, 2020c
Layout: M. Engelbauer, December 2020
© EuroGeographics for the administrative boundaries
© ESA Climate Change Initiative - Land Cover led by UCLouvain (2017)

Source: Own illustration

In 2000, the Ciénaga Grande de Santa Marta and Seaflower Biosphere Reserves were designated. There is neither a national system of biosphere reserves nor a national legal framework that recognizes biosphere reserves. Most biosphere reserves (or at least their core zones) are national nature parks (23:5; 23:10). The Seaflower Biosphere Reserve on the island of San Andrés in the Colombian Caribbean is managed by a very strong local foundation called the Carolina Foundation. The biosphere reserve has legal, administrative and financial autonomy, and a community committee oversees the implementation of activities (23:6; SANTOS-MARTÍNEZ et al., 2013, pp. 11-12).

The MAB Programme in Colombia is managed by the Ministry of Environment with the collaboration of the Colombian National Parks Authority, an administrative unit of the Ministry of Environment. Within the ministry, there are three main groups of actors in the MAB Programme: the directorate of forests, the directorate of seas and coasts and the directorate of international relations, which acts as a mediator. Moreover, the Colombian Commission for UNESCO supports the MAB Programme and has an executive secretary in the Ministry of Foreign Affairs. Thus, the team that supports and manages the MAB Programme in Colombia consists of the Ministry of Foreign Affairs, the Ministry of Environment and the delegation to UNESCO in Paris as a contribution for political and diplomatic support (23:1).

Based in Bogotá, the National Commission for UNESCO consists of five people. However, a MAB National Committee has not yet been created for Colombia. This has been an issue for several years, and there is an ongoing discussion about whether create an exclusive national committee for biosphere reserves or take advantage of existing structures within the government. So far, responsibility for biosphere reserves has rested with the Ministry of Environment, but it is challenging to adapt all relevant information for the periodic reviews and conduct diplomatic negotiations with the MAB ICC. Therefore, it is currently a priority for Colombia to create a MAB National Committee (23:4).

There is an initiative to designate a new biosphere reserve in Chocó next to the Pacific, which has support from UNESCO and the government of Flanders (23:6). Chocó is home to indigenous and native communities and is one of the most biodiverse areas in the world; it has also been affected by the armed conflict in Colombia. Currently, community consultations are taking place to ensure that the zonation is performed correctly and accepted by all communities (23:13). The plan is to prepare the application by 2020. Thus, the creation of a MAB National Committee and the designation of a biosphere reserve in Chocó are currently the main priorities in Colombia, in addition to strengthening existing biosphere reserves (23:12).

5.5.2 Honduras

Honduras has been involved in the MAB Programme since its inception in the 1970s. In 1981,[1] the Río Plátano Biosphere Reserve was designated (22:1). The first-generation biosphere reserve harbours tropical moist forest and wet forest and covers important coastal marine areas consisting of mangrove forest, lagoons and coral reefs in the eastern part of Honduras, near the border with Nicaragua (UNESCO, 2020c). In the first decade of the 21st century, Honduras promoted the designation of new biosphere reserves. Between 2011 and 2017, three more biosphere reserves were designated: Trifinio-Fraternidad; Cacique Lempira, Senor de las Montanas; and San Marcos de Colón. Thus, Honduras currently has four biosphere reserves (see Figure 32; 22:2).

Within the MAB Programme, the Trifinio-Fraternidad Biosphere Reserve has been a model of transboundary cooperation between three countries: Honduras, Guatemala and El Salvador. It was the first transboundary biosphere reserve in Central America and represents a major contribution to the Mesoamerican Biological Corridor. The predominant ecosystems are cloud forest, subtropical dry forest, pine-oak forest and wetlands such as the Laguna Guija, which is a Ramsar site with more than 45 globally threatened species. The area also contains many archaeological sites from the Maya civilization. In 2016, 80 additional municipalities joined the biosphere reserve in an extension of the site. As a result, Honduras now hosts the largest portion of the Trifinio-Fraternidad Biosphere Reserve (22:3; UNESCO, 2020c).

[1] In the interview, 1981 was mentioned as the year of designation, the new UNESCO biosphere reserve database as of 2020 states 1979 as the year of designation, and the old UNESCO biosphere reserve database from 2011 indicates designation on 13 July 1980.

Figure 32: UNESCO Biosphere Reserves in Honduras

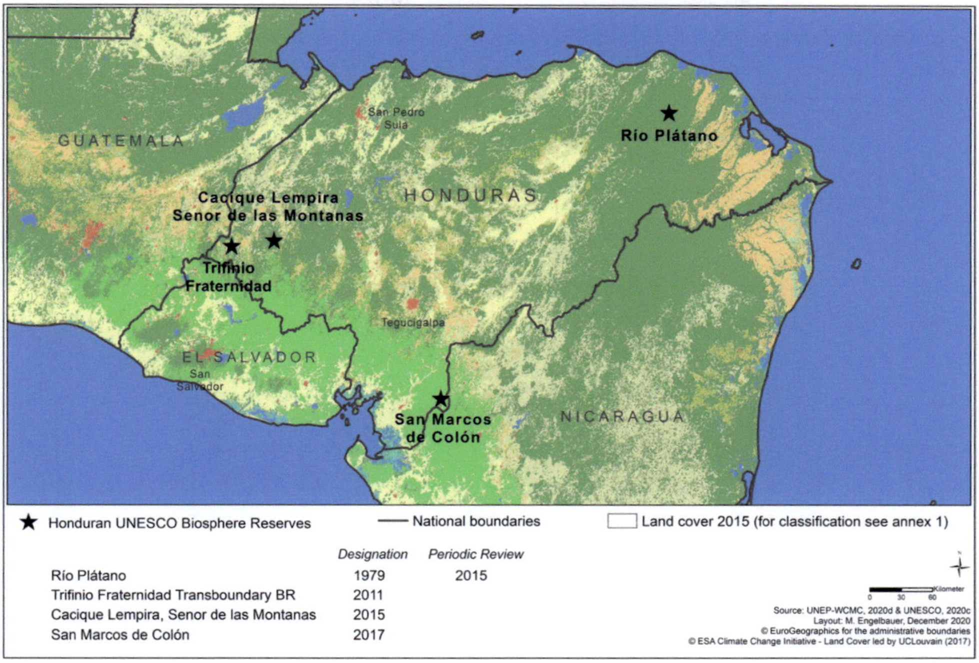

Source: Own illustration

However, Honduras does not have a MAB National Committee. In recent years, there have been meetings and discussions with different authorities to create a MAB National Committee to better promote the network of biosphere reserves in Honduras (22:4). The core zones of biosphere reserves in Honduras are mainly national parks and therefore managed by local offices of the Institute for Forest Conservation and Development, Protected Areas and Wildlife. They work with the Department of Natural Resources, Mines and the Environment (22:8). A MAB National Committee was established in 2011 or 2012 with the participation of the national UNESCO Commission, both abovementioned national authorities and the Ministry of Foreign Affairs. However, only a memorandum of understanding was produced; this facilitated more organized work, but the committee was never effectively established (22:12). During the next regional conference of biosphere reserves in the Americas in Montevideo, it will be important to standardize a model for MAB National Committees in Latin American countries, seek assistance to establish up and learn about best practices. Synergies between Latin American countries are crucial for Honduras (22:13). This is also reflected in working groups within UNESCO. Every month, ambassadors and their staff meet at a plenary session of the Group of Latin American and Caribbean Countries (GRULAC) in UNESCO in Paris to achieve consensus on activities and elections for the MAB Programme. One of these joint decisions resulted in Honduras serving as a member of the MAB ICC and the MAB Bureau from 2018 to 2020 (22:5).

A strategic priority to designate new biosphere reserves in Honduras is to foster the interconnectivity of the Mesoamerican Biological Corridor and establish a transboundary biosphere reserve on the Atlantic coast. There is currently a discussion between authorities in the Gulf of Fonseca, which is shared with Nicaragua and El Salvador, to establish an important biosphere reserve that encompasses crucial mangrove forests (22:14).

6 Results

6.1 Greatest challenges related to the World Network of Biosphere Reserves

In all the interviews, participants were asked about the greatest current challenges in the World Network of Biosphere Reserves and the MAB Programme. The word cloud in Figure 33 shows an abstraction of gathered information and categorisation into keywords for visualisation. It is clear that funding is the greatest challenge with regard to both UNESCO's management of the MAB Programme and the biosphere reserves. After the Seville Biosphere Reserve World Conference in 1995, the shift from a scientific focus on nature conservation to model regions for sustainable development still poses major challenges for biosphere reserves worldwide. These are strongly linked to the implementation of and adaptation to the Exit Strategy introduced in 2013, which forced biosphere reserves to make quality improvements and comply with the minimum requirements of the Statutory Framework. There were also implementation issues related to the zonation and legislation of the three predefined zones: the core, buffer and transition areas.

Another frequently mentioned challenge was a lack of recognition of the value of biosphere reserves by international organizations, national governments and authorities, as well as low acceptance from local populations. The last category of challenges was the naming of biosphere reserves. Often, the word 'reserve' was considered problematic, as it evoked a category of protected areas with strong restrictions for local populations.

Figure 33: Word cloud of greatest challenges in the World Network of Biosphere Reserves identified by interviewed experts

Source: Own illustration using wordart.com

Funding

80% of interviewed experts stated that one of the greatest challenges for the World Network of Biosphere Reserves was funding at the local, national and international levels. Two interviewees highlighted that, whilst the number of UNESCO Biosphere Reserves has increased over time and the MAB Programme plays a new and important role in the SDGs, the MAB Secretariat's budget has decreased (1:2; 31:1). Other significant challenges include keeping track of the growing number of periodic reviews, weighing recommendations provided by the MAB ICC and supporting applications for new sites with fewer programme officers at the MAB Secretariat (31:3; 12:26). So far, this has been possible through the hard work of dedicated people, but a long-term solution is missing (8:17; 1:2).

In addition, at the local and national levels, financial investments are necessary to successfully manage biosphere reserves. The nomination process itself requires funding, as the participatory process is particularly time-consuming and therefore expensive (19:3). After the designation, funding is needed to address conservation, sustainable development and work with local communities (18:2; 9:2). Despite the many demands on biosphere reserves, they do not receive more financial resources (27:5). Since the label 'biosphere reserve' is not accompanied by general funding, administrators are dependent not only on funding from federal states but also on the fundraising of third-party funds using their label (13:12; 26:14). Therefore, it is important to not only formulate objectives and the map but also a resilient governance system with necessary funding in place; furthermore, this should take place not only during the compilation of the technical dossier but also the nomination process for a biosphere reserve (18:26; 1:34).

Paradigm shift since Seville conference and implementation of the Exit Strategy

In the 1970s, the first biosphere reserves were rooted in nature conservation and scientific research. In the 1990s, it became clear that they would not work without the participation of and a level of ownership from local populations. Most activities in biosphere reserves in the 1970s and 1980s were devoted to scientific research on topics such as genetic resources, the production of biomass or the ecological stability of forests, and the publication of results in scientific articles and reports (2:18, 1:1). However, during the Seville conference in 1995, the framework for biosphere reserves was changed, and they became more of a development concept (19:1). Thus, most work focused on the utilization and management of natural resources and how local populations could profit from the ecosystem services provided by biosphere reserves (2:19). As a result, biosphere reserves in some countries still focus on nature conservation, whilst others focus on sustainable development and the relationship between humans and the environment (5:1; 13:1).

A major challenge is to transform pre-Seville biosphere reserves by involving all stakeholders and finding a balance between conservation and sustainable development (1:1). There were different opinions about how strictly the guidelines of the Exit Strategy should be implemented. In particular, there were differences of opinion about the deadline of implementation. Whilst some MAB delegates from European countries demanded a stricter implementation in only two years, MAB delegates from many countries in Latin America and Africa wanted a longer period of time (18:1; 3:17). One interviewee mentioned that 'not everything can be done within two years. (…) So it should be acceptable if the government says it will improve in the next five years' (2:10). In addition, the proposed name of the Exit Strategy was the subject of much discussion, as a strategy should aim to achieve some positive impact, not an exit (2:10). Section 6.2 addresses this key challenge in more detail.

Legislation

A major challenge in the nomination process for new UNESCO Biosphere Reserves is that the entire process must be explicitly confirmed to have been participatory. Thus, the state, the province, the mayor and all local authorities must prove that they agreed with the designation (19:6). In addition, core zones must be legally protected areas, as established in the Statutory Framework. Buffer zones should be placed under national designation, or at least be geographically clearly defined as a protected area. However, transition zones are not necessarily covered by any legislation (1:6). Thus, there is no clear requirement from UNESCO that biosphere reserves as a whole must be laid down in legislation (25:23). In some countries, there are opportunities to use existing legislation to implement biosphere reserves and assign legal status to their different areas. In other countries, the entire biosphere reserve becomes a protected area (13:21).

Weighing the advantages and disadvantages of integrating biosphere reserves into national legislation is a long discussion. On the one hand, doing so provides structure and regulates responsibilities, which are then associated with financial and human resources. Moreover, responsibility for biosphere reserves is usually allocated to a specific department within the government, usually the Ministry of Environment; this is important to ensure accountability (27:7; 27:8). With fewer regulations, biosphere reserves become a societal negotiation process that leads to greater heterogeneity (27:10). On the other hand, classifying entire biosphere reserves as protected areas can be a disadvantage when communicating with local populations, especially if they do not yet understand the biosphere reserve concept and have already had negative experiences with the designation of national parks or other natural protected areas. In such cases, they usually do not want a new protected area (27:9).

In France, Germany and Spain, biosphere reserves are an independent protected area category and thus part of national legislation on nature conservation. In Estonia, Korea, Russia and Slovakia they are mentioned in legal acts, but not as a separate category of protected areas. However, the vast majority of countries have not incorporated biosphere reserves into their legal frameworks. In the 21 case study countries presented in this thesis, only three countries have explicitly included biosphere reserves as protected areas in their national legislation. Three other countries have included biosphere reserves in other legal frameworks but without specifically defining them. Nevertheless, 14 out of 21 countries in this study have not included biosphere reserves in a national legal framework (see Table 12).

Table 12: Overview of case study countries that have explicitly included biosphere reserves as protected areas in their national legislation

Country	Biosphere reserves included in national legislation
Australia	No
Austria	No
Bulgaria	No
China	No
Colombia	No
Estonia	No, but mentioned in Act of Sustainable Development
France	Yes, since 2016 in the law on biodiversity regrowth, nature and landscapes
Germany	Yes, in Federal Act for the Protection of Nature as a distinct category of protected area
Ghana	No
Honduras	No
Indonesia	No, but intended
Japan	No
Kenya	No
Republic of Korea	No, but mentioned in Environmental Conservation Act
Russian Federation	No, but mentioned in Law on Specially Protected Natural Areas
Slovakia	No, but mentioned as territories of international importance
Spain	Yes, in Law 42 of 2007 on Natural Heritage and Biodiversity
Sudan	No
Sweden	No
United Kingdom	No
United States of America	No

Source: Own illustration and data

Awareness and recognition of biosphere reserve concept

In addition to more technical challenges in the implementation of the biosphere reserve concept, the interviewed experts raised many issues with general awareness and recognition of the World Network of Biosphere Reserves. Despite the nearly 50-year history of biosphere reserves (including 25 years in a modern form), the concept still lacks political understanding in many countries. In many cases, few people in politics, ministries and national agencies know about the instrument despite all of UNESCO's strategies and initiatives (25:6; 25:8; 24:2). So far, biosphere reserves tend to be purely part of their research or environmental policies and have low political priority (25:12).

Therefore, the World Network of Biosphere Reserves needs greater awareness at the national, regional and local levels. However, the need for global awareness may be higher (1:38). A challenge for the MAB Programme is to provide more support and advice to national authorities to encourage more national stakeholders to become involved in the global network (8:19). It is not enough for individual countries such as Germany, Spain and South Korea to commit to the MAB Programme. More action is needed at the international political level, and the circle of nations that strategically adopt the issue of biosphere reserves in their development cooperation must be expanded to the 10 large donor countries. At a global level, this would compensate for a lack of a strong coordination efforts at UNESCO in Paris (25:12).

Internationally and nationally, a major issue is a lack of recognition for UNESCO Biosphere Reserves compared to other international protected area labels. Some people have heard of biosphere reserves but do not understand their purposes and benefits, aside from designation. By contrast, UNESCO World Heritage Sites are simple to understand because they solely focus on protection, whilst Global Geoparks generally focus on protection and tourism development. However, UNESCO Biosphere Reserves encompass education, training, nature conservation and sustainable development at one site. Thus, people often have difficulty understanding the concept (12:1). In the United States of America, for example, many people know what an UNESCO World Heritage Site is, but most have no idea what biosphere reserves are (28:3).

Another challenge is ensuring the recognition and acceptance of biosphere reserves by local populations. Especially in first-generation biosphere reserves, the integration of local communities in often old-fashioned protected areas needs to be improved as well as the general knowledge of every farmer and citizen that they live inside a biosphere reserve, remains a major challenge (28:14; 20:1).

In addition, the presentation and communication of biosphere reserves must be further developed to position them as sites for sustainable development. So far, narratives and wording on biosphere reserves in publications and on websites often convey the idea of protected areas. However, it is essential for biosphere reserves to have a multi-departmental function, as is often the case in practice (25:26). One challenge is that many staff members who are directly involved in the work of biosphere reserves are often biologists or forestry experts who lack in-depth communication skills. Although there is important work being accomplished at biosphere reserves, it is very difficult to communicate this to surrounding communities to enable them to better understand the purpose of biosphere reserves (10:1). This requires additional human resources, particularly staff with the skills to engage in participatory outreach and dialogue with surrounding people and communities (31:11).

Another issue is the handling and lack of communication about the huge amounts of data on biosphere reserve activities that are submitted to the MAB Secretariat, most of which are now digital. However, due to a lack of time, they have not been made available despite UNESCO's open access policy (1:23). Yet, these data could be used to promote biosphere reserves and serve as a major resource for research (1:37).

Naming

Out of 21 case study countries examined in this thesis, only half use the term 'biosphere reserve' (see Table 13). Particularly in English-speaking and European countries, the term 'reserve' causes significant problems with local populations. In Australia, for example, the term 'reserve' tends to stall discussions (particularly with rural inhabitants), as it connotes a complete lack of access and many local people do not want any more restrictive protected areas. Therefore, Australia only uses the word 'biosphere' (9:2; 14:4). In the United Kingdom, biosphere reserves are also simply called 'biospheres' (1:25). In Germany, the situation varies; depending on the federal state, they may be called 'biosphere reserves', 'biosphere areas' or 'biospheres'.

Table 13: Overview of the usage of the term biosphere reserve in case study countries

Country	Usage of 'biosphere reserve'	Alternative wording
Australia	No	Biosphere
Austria	No	Biosphere Park
Bulgaria	No	Biosphere Park
China	Yes	
Colombia	Yes	
Estonia	No	Biosphere Area
France	Partly	Biosphere
Germany	Partly	Biosphere Area, Biosphere, Biosphere Region
Ghana	Yes	
Honduras	Yes	
Indonesia	Yes	
Japan	No	UNESCO Eco Parks
Kenya	Yes	
Republic of Korea	Yes	
Russian Federation	Yes	
Spain	Yes	
Sudan	Yes	
Sweden	No	Biosphere Area
United Kingdom	No	Biosphere
United States of America	No	Biosphere

Source: Own illustration and data

In Austria, the term 'reserve' reminds many people of an Indian reservation and an area in which people are put under a glass bell; therefore, the sites are called 'biosphere parks' (18:23). In Bulgaria, the protected area category 'strict nature reserve' is included in national legislation and is the most restrictive category in the country. Therefore, many Bulgarians link the word 'reserve' with regimes and limitations. To avoid raising concerns and reduce the likelihood of local populations refusing to join the MAB Programme, post-Seville sites are called 'biosphere parks' (30:14; 30:15; 15:16). In France, the term 'biosphere reserve' is used at the national level; however, at the local level, people are free to use what they want. Many use 'the biosphere' (24:7). The term 'reserve' does not have very positive associations in Sweden, as it often refers to a place that people are not allowed to be in. Thus, biosphere reserves are called 'biosphere areas' or simply 'biospheres' in Sweden (10:7).

By contrast, naming is more diverse is Asia. In China, there are no issues with the term 'biosphere reserves' because nature reserves are protected under national legislation and many also use the word 'reserve' (7:14; 12:3). The Republic of Korea still uses the term 'biosphere reserves', but there have been ongoing discussions on changing it. The word 'reserve' connotes strict protection, and many local people are concerned about their properties being restricted in the future (12:2). Another difficulty is that the translation of 'biosphere reserve' in Korean is quite long. For example, 'Jeju Island' in Korean is three characters but 'biosphere reserve' is seven characters (12:8). Perhaps the greatest deviation from 'biosphere reserves' is in Japan, where the sites are called 'UNESCO Eco Parks'. Because the term 'biosphere reserve' is too academic and not commonly known in Japanese society, the MAB National Committee decided to change the name in 2010 to increase popular interest in the concept (21:8).

In the case study countries from Africa, the Arab States or Latin America and the Caribbean, there were fewer problems or deviations from the term 'biosphere reserves'. This is mainly due to the fact that the word 'reserve' occurred more often in the protected area categories of these countries and that there were national reserves and game reserves in addition to biosphere reserves (8:9; 23:11; 9:11; 3:6; 20:10).

UNESCO has discussed the naming of biosphere reserves for decades and is relatively generous in accepting different naming (18:23). The MAB Programme allows member states to use their own terms in relation to biosphere reserves (15:16). However, some critics have called for standardization of the name (1:25). One option would be to call them 'Man and the Biosphere Sites' or 'MAB Sites' for short (21:16).

6.2 Quality improvement strategies: the Exit Strategy and the Process of Excellence

The MAB Programme began in 1971, and the first UNESCO Biosphere Reserves were designated in 1976. The concept was novel and unique at the time, and there was a huge rush to designate sites. Some countries, such as the Soviet Union and the United States of America nominated a large number of sites from the onset (19:10.1). However, the concept was not as defined as it is today, and UNESCO did not have very strict rules or guidelines on how to run the MAB Programme (7:8). At the time, biosphere reserves were viewed as a type of protected area. However, from 1995 onwards, the concept considerably changed. In addition, it became clear that protected areas should be administered by the countries themselves, not UNESCO. However, the integrative approach of sustainable development was new and interesting. Therefore, UNESCO member states decided to significantly change the concept of biosphere reserves, which had a substantial impact on existing and future sites (19:10.1).

It is important to note that UNESCO is the only UN organization that designates certain geographical areas, and biosphere reserves are the only concept that encompasses three different zones and combines protection and land use. World Heritage Sites only have core and buffer zones, whilst Global Geoparks do not have any core zones or protected areas. Thus, biosphere reserves are unique but also often difficult to manage (19:4, 19:5).

After the Seville conference in 1995, several European countries submitted innovative proposals for second-generation biosphere reserves. The discrepancies between different sites became increasingly obvious (13:17). At the Seville+5 meeting in Pamplona in 2000, it was clear that there were severe issues with some biosphere reserves, which did not conform to the Statutory Framework (13:16). Since the adoption of the MAP, it has been clear that biosphere reserves that cannot meet the criteria of the Seville Strategy must be withdrawn from the global network (25:31).

Current biosphere reserves in the World Network of Biosphere Reserves are commonly divided into two groups: first-generation biosphere reserves (up to 1995) and second-generation biosphere reserves (1995 onwards). In the aftermath of the MAP, the MAB ICC became increasingly concerned about including first-generation biosphere reserves must also be included in the new concept to ensure that all sites in the World Network of Biosphere Reserves function according to the Seville Strategy (19:10.2).

In the early years of the Seville Strategy, some delegations did not take the periodic review process seriously because they knew that the director-general of UNESCO would not sign letters of withdrawal (1:16). However, a clear line was drawn in 2013, and an analysis was made of how many sites were concerned. A preliminary evaluation revealed that 270 biosphere reserves in 85 countries were affected and did not function as they should (19:10.2).

At the time, there was also a significant shift in the members in the MAB ICC. Whilst attendees at the MAB ICC meetings had mainly consisted of older men from the natural sciences in the past, this changed with the arrival of younger attendees from countries that had already established new post-Seville biosphere reserves (1:16). The preliminary evaluation and the shift in the members of the MAB ICC led to the MAB Secretariat's decision to send three more reminders to biosphere reserves that had not fulfilled the criteria, never submitted a periodic review and never responded to the MAB ICC's recommendations; if they still did not act, the designation would be withdrawn. This resolution was very ambitiously formulated in 2013 and marked the beginning of the so-called Exit Strategy (25:32).

Although the letters were sent out, it remains unknown who actually received them. It is unclear whether they were forwarded to the managers of biosphere reserves or stalled at the MAB focal points and the countries' permanent representation in Paris (25:32; 27:38). The major decision on how to deal with the affected biosphere reserves from the Exit

Strategy was intended to be made at the UNESCO World Conference in Lima in 2016, but more time was given because it was unclear who had actually received the letters (25:32).

There was also a long discussion about what would happen to biosphere reserves affected by the Exit Strategy. Some states demanded that non-functioning biosphere reserves be removed from the World Network, whilst others called for help to be given to affected sites to ensure that all biosphere reserves conformed to the Statutory Framework. Of the 270 biosphere reserves affected by the Exit Strategy, only one third remained by 2017. The rest had either successfully completed the adaptation process or voluntarily withdrawn from the World Network of Biosphere Reserves (19:10.2).

For example, Germany, Norway and the United Kingdom withdrew biosphere reserves because they had good research areas and protected areas that had been designated as first-generation biosphere reserves but could not fulfil the sustainable development function (5:4; 16:14). In Germany, the Bavarian Forest was withdrawn as a biosphere reserve because the national park administration said that it could not develop a transition zone and the local population strongly opposed its potential expansion (5:3).

However, an issue that received relatively little attention in the implementation of the Exit Strategy was the MAB ICC's process of issuing designation withdrawals for UNESCO Biosphere Reserves, especially if many biosphere reserves would be affected in 30 or 40 countries at the same time. Since UNESCO is a member state organization, politics are inevitably involved, which means that some countries require more diplomacy to get them withdraw sites due to their position within the organization. Even if countries do not sit on the MAB ICC, they probably have close allies there; as it is an intergovernmental body, ambassadors also attend meetings for important decisions. Thus, the question remained of how to solve withdrawals pragmatically to ensure the effectiveness of the Exit Strategy (25:33; 13:1).

Whilst the discussion was relatively peaceful over the years, it became heated during the MAB ICC meeting in 2017. The European countries had proposed to the MAB Council to exclude 85 biosphere reserves and withdraw their designation. Under the leadership of GRULAC and a few African countries, the proposal was voted down; many argued that the sites should remain in the World Network of Biosphere Reserves and that members should continue to work towards improving them together or let them voluntarily withdraw (19:10.2; 18:21; 7:8; 8:6; 23:2). Under negotiations of the French chair of the MAB ICC and Council, Didier Babin, compromises were negotiated. It was decided that the biosphere reserves still concerned would be excluded from the World Network; however, this would only come into force two years later and thus allow enough time to prepare a new application and prevent loss of the designation. However, submitting additional documents was not enough; an entirely new application would have to be submitted. Thus, there would be a clear increase in the quality of the World Network of Biosphere Reserves over the next few years, even without the immediate withdrawal of biosphere reserves (25:33; 18:21). During the interviews, Babin commented that this compromise was only possible because there was significant trust within MAB Programme:

> 'It's one of the main differences in the biosphere reserve programme in comparison with the World Heritage Convention. We trust each other, we work in the network,

> we have no outsiders coming in and saying you are not following things. No, it's between us and I think this is interesting'. (Didier Babin, 2018)

If biosphere reserves lacked funding and had a difficult social situation, they had to rely on the solidarity of other member states (19:10.3). In particular, countries that wanted to stay in the network should be given more time and supported to improve their biosphere reserves to meet the standards (7:25; 2:10; 24:15). At the end of the MAC ICC meeting, the Exit Strategy was renamed to the Process of Excellence, with a focus on more general capacity building (23:2).

During the many bilateral consultations at the MAB ICC meeting in 2017, the ambassadors of the affected countries made it clear that the issue of adapting biosphere reserves to the Seville Strategy was recognized and that it had been put on the back burner until then. They realized that zoning could only be achieved as part of a larger process. However, it was also important to understand that the process was not about indiscriminately excluding and punishing some biosphere reserves but rather permanently strengthening the other 90% of biosphere reserves and the entire MAB Programme, especially in such a difficult financial situation for UNESCO (27:36).

Success with the implementation of the Exit Strategy widely varied. Some sites made real progress and instituted significant changes. They changed area boundaries, included local communities and started interesting new initiatives. Differences in the success and failure of the Exit Strategy's implementation cannot be attributed to individual countries, as there were notable differences even within countries. This is exemplified by the United States of America and Bulgaria (13:18; 13:19).

A major point of criticism in the implementation of the Exit Strategy was the speed with which new transition zones were designated. In response to the Exit Strategy, transition zones were created around first-generation biosphere reserves in many parts of the world. Given the speed with which these were created, it can be deduced that either local populations were not as involved as they should have been or were promised many benefits, with a somewhat doubtful prognosis (25:19).

In addition to spatial changes in the concept of biosphere reserves and the adaptation of first-generation biosphere reserves, another critical challenge was the requirement to include a much wider range of stakeholders and implications for power relations. Managers who were responsible for protected core zones had to share or even relinquish power to include more people and organizations in the management of the biosphere reserve. A major disadvantage of this approach is that then management is no longer a single party's responsibility but everyone's responsibility, which raises the question of who should pay for it. This is a global issue and creates more tension in the implementation of post-Seville biosphere reserves (1:7).

However, according to the interviewed experts, the greatest criticism of the Exit Strategy was its name. Many felt that the wording 'exit' had negative connotations, as any strategy should aim to improve a situation and achieve positive outcomes, not an exit (2:9; 2:10; 1:18; 12:17; 9:4; 16:14; 10:18). Thus, renaming the Exit Strategy to the Process of Excellence was crucial, as it signalled a quality improvement approach (14:25; 8:17; 10:18).

Therefore, all interviewed experts agreed that quality improvement strategies were important. After some countries had not sent 10-year periodic reports for decades, the Exit

Strategy was fundamental to improving the credibility of the MAB Programme (26:16; 28:2; 4:2), which was urgently needed (31:13). The reality of two quality levels in one programme was no longer tenable, both within and outside UNESCO, as the MAB Programme could not be communicated in a serious way. Although gaps in quality remain between biosphere reserves after the Exit Strategy, it was the best strategy for adapting old and non-functioning biosphere reserves to the new standards (18:31) or encouraging member states to voluntarily withdraw biosphere reserves from the network (21:16). Thus, the remaining biosphere reserves are now in a different stage with local acceptance and recognition (28:2). Furthermore, three national focal points for biosphere reserves in different countries said that the Exit Strategy was also very helpful for exerting pressure on biosphere reserves and thus revitalizing the biosphere reserve network in their country (24:14; 21:5; 26:16).

6.3 Implementation and improvement of the periodic review system

Ten-year vs. five-year reporting cycle

During the interviews, one of the greatest points of discussion with regard to periodic reviews was the 10-year cycle for compiling and submitting evaluations to UNESCO. The question was whether it was viable to wait 10 years to see what happened in the biosphere reserves and the technical and financial feasibility of shortening the reporting cycle. Eleven experts clearly favoured a shorter reporting cycle (e.g. five years), whilst four experts wanted to keep the 10-year reporting cycle.

Some believed that the 10-year reporting cycle was appropriate, as it is difficult to evaluate anything in less time (14:21; 31:9). This is particularly true of biosphere reserves, as they represent a long-term commitment and involve lengthy collaboration processes. Therefore, effects often take a long time to appear. For example, the UNESCO designation of one biosphere reserve as a model region for sustainable development actually changed the entire municipality's approach to transport. The municipality has the idea to use locally produced, renewable energy for all trucks in the area. Thus, the biosphere reserve inspired a new vision for the municipality; from this perspective, 10 years is an appropriate reporting cycle (10:14). Furthermore, the aim of a periodic review is to verify whether a biosphere reserve fulfils the criteria of the Statutory Framework and still meets objectives 10, 20 or 30 years after its designation (24:10).

On the other hand, many other of the interviewed experts believed that a 10-year cycle for periodic reviews was too infrequent (21:10; 6:13; 1:20). Much can happen within this period, and the structure of biosphere reserves may have changed to some extent. Biosphere reserves are not fixed; they are a living concept. A reserve can become larger or smaller, fewer or more people can live in it and local authorities and managing directors can change. These developments must be communicated to UNESCO to determine whether the biosphere reserve still complies with the current concept (19:7; 31:9). Thus, with more and more pressure to protect areas around the world, it is necessary to determine whether a site is in increased danger and discuss how to troubleshoot any issues that emerge. Thus, waiting 10 years may be too long (29:3).

Indeed, the evaluation period for biosphere reserves is the longest of all internationally designated areas (1:20). For World Heritage Sites, the reporting cycle takes place every six years. Thus, it may be preferable to change the reporting cycle for biosphere reserves to every five years or the same length as World Heritage Sites (12:12; 21:10; 7:7; 26:17). However, the main challenge with a five- or six-year reporting cycle is that they entail much more work for biosphere reserve staff and supervisory bodies. Thus, more technical and financial resources must be allocated to staff for additional reporting (7:5; 28:18; 26:17; 1:20).

One proposal is to conduct a five-year supplementary review with a simpler questionnaire to provide a snapshot of the status of each biosphere reserve, with some basic facts about the health of the local ecosystem (26:7; 3:7). This could be a supporting process for the main 10-year periodic review and allow discussions to be held with all stakeholders at the midpoint of each cycle (31:9).

The periodic review process could also be revised and streamlined. If one assumes that all biosphere reserves would more or less conform to the Statutory Framework criteria after the Exit Strategy and the Process of Excellence and Enhancement, then the periodic review itself could become less labour-intensive (1:23). Currently, it consists of a long and complex form; filling it in and collecting all the necessary information take significant time and work (3:7). On the one hand, filling out the form is very useful as an information-gathering process. However, if the main goal of the periodic review process is to meet the criteria of the Statutory Framework, only a small part of the form actually focuses on this (1:22). Thus, a simpler version of the questionnaire with rubrics that people can assign scores to may facilitate more regular updates; this is difficult to do with the current form (13:13).

Zonation

During the implementation of the Exit Strategy, it also became apparent that many countries had substantial difficulties with zonation. For example, in the United States of America, there has historically been an issue with the concentric structure of biosphere reserves, which involve a protected core zone, a buffer zone and a transition area. The word 'buffer' is very sensitive due to strong private property rights in the United States of America (28:1), as landowners do not want to give up control of their lands (13:3). However, since the core zone must be surrounded by a buffer zone, parts of the core zone are sometimes transformed into the buffer zone because there is simply no space or opportunities to put them elsewhere (13:3).

In its periodic review, the United States of America made a proposal to rename the buffer zone to 'area of transition' or 'joint-management area' and the transition zone to 'area of cooperation' (28:1). The term 'transition zone' has an old-school meaning; for example, 'sustainability area' would be better (21:25).

Self-evaluation vs. external evaluation

Periodic reviews for biosphere reserves are self-evaluations conducted by member states. By contrast, an external figure from IUCN evaluates World Heritage Sites (19:7). Thus, the findings from periodic review reports must be taken at face value (1:35). Many people believe that self-evaluations are not a serious system; however, they ultimately depend on implementation in different countries. Some member states take periodic reviews seriously and therefore have an effective system (24:12). However, it would be ideal to replace self-evaluations with a peer review system. In the MAB IAC's experience, documents and information in periodic reviews do not always represent the real situation on the ground. In a voluntary peer review system, another biosphere reserve in the same country or another country in the same region would also examine the periodic review. Thus, biosphere reserves could still conduct their own reviews, but another biosphere reserve would also check it over. This could be beneficial to both parties and potentially increase twinning arrangements in the network (10:12; 21:11). The voluntary peer review should be seen as a strength and an added value when handing in the periodic review to the MAB IAC and ICC (10:13).

Another option would be to have an external evaluation team examine each biosphere reserve and make recommendations on what has happened, what is going well and what gaps should be filled. This would be a more comparable and objective approach, but it is relatively expensive (18:18). Since UNESCO and the MAB Programme have limited funds, there has not yet been an initiative to institute a similar evaluation system as the World Heritage Sites (26:17).

Increased workload for the MAB IAC

One of the consequences of the growing number of UNESCO Biosphere Reserves worldwide is increased workload for the MAB IAC and the MAB Secretariat (16:11). Apart from the evaluation of new designations, the assessment of substantial changes to existing biosphere reserves and the screening of periodic reviews also falls under the MAB IAC's responsibilities. With the Exit Strategy, it must also respond to questions and provide recommendations to countries to on making certain changes or withdrawing sites from the network (13:5; 12:18). It is particularly difficult for the MAB Secretariat to keep oversight and for the MAB IAC to review all documents (1:35). In addition, MAB IAC meetings are very short. Thus, it is very challenging even for experienced experts to carefully work through the periodic reviews (12:14; 16:11). Moreover, work on the Advisory Committee is voluntary, and most members must take a few days off from other commitments for this purpose. It takes dedication to engage in such work, and this would increase if the reporting cycle is shortened in the future (13:20).

Due to a high number of applications, not all members of the Advisory Board can review all applications. This is why the MAB IAC divided itself into regional groups to receive the applications for evaluation (18:12). In principle, these groups are completely separate, but some exchange is allowed in case experts have experience in one of the countries or a specific biosphere reserve under consideration (13:24). The composition of these groups can widely vary. Some contain more social scientists, whilst others may have more conservationists or biologists. Therefore, they can produce different assessments about zonation and boundaries between possible buffer zones and core zones. This results in different quality standards (18:12). With more members in the IAC, there is a need to develop stricter regulations and technical guidelines to provide clear standards for the evaluation of applications (12:18).

Regarding improvements, there is a debate about strengthening communication and exchange possibilities between the political and technical apparatuses of the MAB Programme (i.e. between the MAB ICC and the MAB IAC), as there are many issues with dialogue between these bodies. For example, no members of the scientific committee participate in MAB ICC meetings. The only intermediary is the MAB Secretariat, which shares news from the IAC with the ICC, and vice versa. However, there is never a direct connection between the two bodies. Thus, it is recommended to hold a midterm meeting with the MAB ICC and the MAB IAC to foster synergies and initiate discussions about restructuring the MAB polices and regulations used by the IAC. Not all information should only go through the MAB Secretariat (8:20).

Regional networks for enhanced quality management

Regarding quality management, it is important to note that states themselves make a commitment to overseeing biosphere reserves to UNESCO. States are members of UNESCO and participate in the MAB Programme. If a state decides that it wants its sites to be internationally recognized as UNESCO Biosphere Reserve, then it must fulfil certain obligations in terms of quality management and periodic reviews (27:32). However, there is flexibility with regard to the specific party that oversees quality assurance. In some countries, the UNESCO Commission is responsible for this; in others, it is the MAB National Committee or state authorities (e.g. nature conservation authorities). This must be decided by the countries themselves and not imposed by UNESCO (27:33; 27:34).

There are several models for conducting periodic reviews. In Canada, for example, the MAB National Committee travels to the biosphere reserve; in Germany, the MAB National Committee comments on and discusses the biosphere reserve's evaluation report onsite (1:20). In China, there is a two-tier system that consists of a pre-review before the formal review. The staff of the MAB Secretariat first visit the biosphere reserve, then an external expert panel evaluates its work. Their suggestions and comments are also translated into English and submitted to UNESCO with the periodic review (7:6). In Sweden, the periodic reviews are conducted by subcontracted external consultants who are supported by the national coordinator for biosphere reserves (16:10).

Currently, it would not be viable to establish more quality management measures at the UNESCO level because the UNESCO system is already far removed from events on the ground and has severe financial limitations. Instead, additional quality assurance measures and evaluation systems must take place at the national or perhaps regional level (27:22). Although UNESCO could provide some resources, it often has more limited financial means than many member states (1:20).

Thus, regional support systems could be an important source of reinforcement (27:39). Two or three countries could join forces (27:28), r regional MAB networks could play a leading role in additional quality management (14:22). For example, periodic reviews could first be submitted at the regional level, then to the MAB Secretariat. In this way, there would be more regional involvement in the global network (1:35). Some members of the regional networks are closer to the biosphere reserves, understand local cultures and speak local languages (1:21). There are already peer-to-peer exchange programmes and training courses for biosphere reserve managers within individual regional networks (21:13). These regional meetings offer platforms to test ideas, find new partners for projects, gain inspiration and learn about one's own work by talking to colleagues in other countries. Thus, members can get to know each other and the different biosphere reserves. These exchanges would certainly benefit a peer review approach to periodic reviews (10:16).

However, this also costs money, which is also an issue for regional networks (1:21; 21:13). Another critical point is whether such exchanges would be possible from a political perspective (18:16). Some countries might not allow nationals from other countries to evaluate and read periodic reviews before they are sent to UNESCO (2:15).

6.4 Interlinkages with other UNESCO labels

World Heritage Sites are outstanding cultural or natural areas, which humanity must not lose. Thus, all efforts must be made to protect these areas. This is why it is also a convention. World Heritage Sites feature a buffer zone around a protected area, which provides protection against threats. There are now over 1,070 World Heritage Sites worldwide, and the network is self-financing. Then, there is the concept of biosphere reserves with an integrative idea. These are model sites, not areas that are outstanding and unique in themselves, but just the opposite. Whilst World Heritage Sites can't be copied, biosphere reserves can and should be copied. This is a clear difference between the concepts. By contrast, Global Geoparks consist of geologically interesting places; they focus less on zoning and more the area as a whole. They do not contain any legally protected zones; therefore, designation is much easier. It is not necessary to ensure the presence of a legally protected core zone, as with biosphere reserves, or worldwide outstanding status, as with the World Heritage Sites. These observations show that the three UNESCO labels for nature conservation are completely different concepts. In some countries, the models are used in a complementary way; in others, they overlap (19:21). For example, Jeju Island in the Republic of Korea has all three UNESCO labels (23:9).

The question of how these three UNESCO labels relate to each other is a controversial one. For example, are they in competition or complementary? Does it make sense to have multidesignated areas with two or even three labels for the same place? How will the diversity of labels be received by citizens? The answers and opinions of the interviewed experts are presented in the following paragraphs. Compared to previous topics presented in this thesis, there was no consensus on these questions.

All the concepts have their raison d'être and are complementary. Although there can be overlaps, they should be avoided. The world is large; the more areas that can be managed under one designation, the better (19:12.2; 23:9). Biosphere reserves are less about beautiful nature conservation areas and more about rural populations. In this context, the concepts can be used in a complementary but sensible way (19:12.2). For example, linking World Heritage Sites and Global Geoparks can make sense if there is a unique geological structure. Geoparks focus on communication and valorisation of the label (27:41). With regard to tourism, for example, Global Geoparks can provide added value by attracting even more attention to the site (8:16). Many people understand that geoparks are distinct from other types of protected areas. It is now important that an additional label is not added to an existing area, but that the actual added value is presented. For example, Global Geoparks could be designated in regions that are not yet involved in regional development and tourism, especially transboundary tourism (25:4). However, geoparks and biosphere reserves are not complementary; the focus must either be on geological features or living nature. Thus, combining these labels can be very tricky (1:28). Some participants said that it would not make sense to conflate the two (27:41; 14:27). They have similar objectives; therefore, geoparks are more in competition with the concept of biosphere reserves (12:20).

However, Biosphere reserves and World Heritage Sites can be complementary, which is evidenced by success stories (1:28). The World Heritage Site designation can provide substantial support for the protection of the biosphere reserve's core zone. In addition, protection of the wider landscape can be integrated through the buffer and transition zones of biosphere reserves (6:18; 6:29; 21:20; 27:40). World Heritage Sites and biosphere reserves also have more opportunities to learn from each other (21:18). The communication of the designations can be complementary. To market a site as a tourist destination, the World Heritage Site label can be used due to its outstanding universal value. However, the biosphere reserve label can be used to promote sustainable development at a site that allows human resource use and supports livelihoods (9:16). For example, the handling of these designations is very clearly defined in Honduras. World Heritage Sites are overseen by a focal point in the Natural Resources Secretariat, and biosphere reserves are overseen by another focal point at the Forest Science Institute. Thus, the labels are complementary, and this way of double designation is considered important for other areas and countries (8:16). This combination of World Heritages Sites and biosphere reserves is the best way of ensuring long-term commitment to maintaining the sites' heritage values and giving them the necessary protection. Biosphere reserves with core zones that are also World Heritages Sites are an ideal working model, but there are not enough of these in the world (14:28).

In the past, natural World Heritage Sites and biosphere reserves were managed by the same secretariat at UNESCO. Then, a decision was made to bring all the World Heritage Site activities, cultural and natural sites under the umbrella of the World Heritage Center to improve visibility because senior staff in UNESCO saw a potential money earner for them. The World Heritage Sites had a much greater political visibility. Unfortunately, this development prevented joint management of biosphere reserves and World Heritage Sites. In addition, joint management could also have promoted exchange in the nomination of WHSs on a possible linkage with biosphere reserves in the buffer and transition zone. However, this is now more difficult with the separation of the MAB Programme and the World Heritage Center within UNESCO (14:39).

Other interviewed experts saw clear competition between the different UNESCO labels. In particular, biosphere reserves have a weak standing; other labels such as World Heritage Sites are perceived more attractively due to the lack of political attention given to biosphere reserves (25:14). In Austria, for example, a biosphere reserve had to be extended due to the Exit Strategy; however, there was concern that there would be many additional restrictions. In discussions with the agency that oversees the protected area, it was decided that the biosphere reserve designation was not important because it was already a World Heritage Site and a national park; these labels were considered to be more important. Thus, the biosphere reserve label was withdrawn (18:30). Another issue is funding. There is competition between different programmes to show that they have more and better sites. In terms of funding, these labels compete with each other, and programmes do not always work together

(13:9; 10:20). It would be easier to explain the UNESCO labels to governments and politicians if there was only one UNESCO designation. However, once explaining the difference between biosphere reserves and geoparks, they enter in a competition of concepts. It would be better if there were not different UNESCO designations within a country (16:17). In addition, there is significant confusion within communities that live around UNESCO-designated areas. For example, one area in Kenya is simultaneously a World Heritage Site, a UNESCO Biosphere Reserve and a RAMSAR site. One week, government stakeholders visit local communities to talk about World Heritage Sites; the next, a different set of stakeholders may talk to them about biosphere reserves. Thus, an upcoming priority would be to facilitate synergies (29:5).

Unlike Global Geoparks, biosphere reserves do not significantly compete with World Heritage Sites because they have different objectives; they each fulfil their own functions and can co-exist without any issues. The greatest competition is between Global Geoparks and biosphere reserves, as their objectives overlap: regional economic development. However, Global Geoparks mainly focus on tourism. In addition, they must have a management plan, it a manager and a participatory body. The same conditions apply to biosphere reserves, but they must cover both tourism and sustainable socio-economic development. Thus, both labels compete for funding and resources. In the past, UNESCO has designated new geoparks in areas that were already biosphere reserves. However, no coordination took place between both programmes (26:19).

Seventeen years ago, the MAB Programme had the opportunity to adequately include the aspect that geoparks embody today. Originally, the idea was to identify and designate a secondary category of biosphere reserves that are much more geologically oriented within the framework of the MAB Programme. However, the MAB ICC decided against this at the time. Then, five years ago, there was an idea to create an additional UNESCO label. Some countries made it very clear that they wanted the Geoparks label (25:28), especially China. China desired multiple labels to boost domestic and international tourism. Thus, World Heritage Sites, biosphere reserves, national parks and geoparks can be found in the same region; sometimes, they even have the same boundaries (14:27). Currently, the existence of three labels is a reality that member states must contend with. This is the nature of international organizations looking for high visibility. If someone is successful with something, such as UNESCO with World Heritage, then other organizations want to copy it as much as possible. This can rarely be prevented. Neither in intergovernmental organizations like UNESCO nor in other non-governmental organizations (25:28).

The concept of Global Geoparks closely resembles the MAB Programme. Many elements from the framework for biosphere reserves have been copied, which raises the question of why certain geoparks do not simply become biosphere reserves. The answer is often simple: it is much easier to obtain the Global Geoparks label (18:29).

At the UNESCO level, a major challenge is that there is no clear management plan for all three nature conservation labels. There should be a mechanism or strategy in place to operate these three different programmes (7:28). In addition, in terms of staffing at UNESCO, there is only a dozen staff members at the MAB Secretariat, who were already looking for more funding. Then, the Global Geoparks programme suddenly fell under the responsibility of the same division, which increased competition for resources (10:22).

In addition, most tourists and locals cannot distinguish between the different UNESCO labels. They have heard of UNESCO and associate it with World Cultural Heritage Sites, but most people do not know that there are also World Natural Heritage Sites and UNESCO Biosphere Reserves (25:29; 27:40). However, the general public may not need to know the difference between the three labels. If they recognized that the region is a UNESCO-designated region and associate it with positive attributes, then this would be a good outcome. Most people know about World Heritage Sites and view them in a positive light. However, they are not familiar with biosphere reserves. Since the latter are a process-oriented category, people do not necessarily need to understand what biosphere reserves are, as long as they know that they are sites where everyone works towards increasing sustainability. If they can understand and be proud of this fact, this is already an achievement (10:23). There is now also increasing use of the label 'UNESCO sites' for all three categories (25:30). It might also be interesting to have a single UNESCO logo for all three labels. Sites could then use these logos more effectively in their marketing to tourists and locals. On the UNESCO homepage, there also needs to be a better explanation of the differences between a geopark and a biosphere reserve (18:9).

Last but not least, the reporting processes for the three UNESCO labels should be harmonized. For example, it would be complicated for one person to manage all three labels in one area. In the same year, they might need to write a report for Global Geoparks, which entails one set of questions, and another for UNESCO Biosphere Reserves, which entails a different set of questions. However, the Statutory Framework specifies a reporting cycle of 10 years for biosphere reserves, while the policy documents for Word Heritage Sites specify a reporting cycle of six years. Geoparks are even more rigorous, with a reporting cycle of two years. Changing the legal basis and conventions of the existing programmes to align the reporting periods is certainly difficult, but a start could be at least to better align the contents of the reports (1:26).

6.5 How many UNESCO Biosphere Reserves will be needed in the world?

The number of biosphere reserves needed in the world is a frequent question, especially in relation to the other UNESCO labels (i.e. World Heritage Sites and the growing Global Geoparks network). The vast majority of experts interviewed still saw growth potential for the World Network of Biosphere Reserves. Some were more vague and expressed a desire for more sites, without giving an exact number. One interviewee stated that more biosphere reserves would mean more citizens learning about sustainable development and nature conservation for future generations (12:23). Adding more biosphere reserves would also increase their visibility and support. With the addition of every new biosphere reserve, more politicians are involved and can pressure parliaments and governments, which is crucial for more visibility and funding (24:20). Other participants were more concrete and envisioned 1,000 (25:40) or even 1,500 to 2,000 biosphere reserves in the World Network (12:23).

However, two out of 31 interviewed experts would like to see fewer biosphere reserves to strengthen the MAB Programme (14:29; 28:22).

At the time of the interviews, there were biosphere reserves in 120 countries (currently, there are biosphere reserves in 131 countries). This means that more than 60 countries do not yet have biosphere reserves. In New Zealand, for example, there are challenges with land allocation. Most of the land that is not directly owned by the state or private individuals is crown land, which complicates the establishment of biosphere reserves by the national government. There are also many small island states, which are often too small to include the necessary zones; therefore, they have not yet designated any biosphere reserves. Compared to the World Heritage Convention, there is no limit of nominations of sites with only one World Heritage Sites per country per year. The limitation for biosphere reserves is that the dossiers must align with the Statutory Framework. In the bottom-up approach to nominating biosphere reserves, the sites have to must also justify themselves to their respective governments: local governments, provincial governments, national governments, the Ministry of Foreign Affairs and the MAB National Committee. Therefore, each country must determine the right number of biosphere reserves for itself. This is part of the innovative UNESCO Biosphere Reserves concept (19:13; 19:14). Some experts perceived the MAB IAC's limited ability to examine periodic review reports as a limitation (21:21; 12:23). Therefore, any expansion of the World Network should accompany the strengthening of UNESCO's MAB Programme and the recruitment of additional experts to work exclusively for the World Network of Biosphere Reserves (23:13).

Regarding the distribution of biosphere reserves and the identification of possible prioritization areas, there were different approaches and perspectives. On the one hand, the natural representativeness of biomes and ecosystems in the World Network must be considered. On the other hand, there is a need for model regions for sustainable development, with a socio-economic focus on the relationship between humans and the environment.

To designate new biosphere reserves, one criterion is the representative ecosystem of the site's biogeographic province. Thus, many ecosystems from the nearly 200 biogeographical provinces in the world should represented in the World Network (12:23). From Central Africa to Southern Africa, for example, there is relatively well-preserved bush forest along the Miombo belt that is not yet represented in a biosphere reserve. There are hundreds of different ecosystems in the Miombo forest that should be conserved and have not yet been adequately recorded. At the same time, some ecosystems in the World Network are overrepresented (25:40). In Ghana, the plan is to represent every ecosystem. So far, the country contains biosphere reserves in forest ecosystems and around natural lakes, but none for dryland and mountain areas. Thus, the focus is not on quantity but the representativeness of ecosystems in Ghana and using the biosphere reserves to educate people about sustainable development (3:14). A similar approach exists in China; the Chinese MAB Secretariat envisions additional biosphere reserves that represent specific ecosystems in the fourth-largest country in the world, which are not yet covered by the network (7:29).

Mapping the world into different categories such as biogeographical provinces represents a typical natural science approach, which generally aligns with the history of the MAB Programme and its origins in the natural sciences (10:25). It was stipulated that biosphere reserves should be ecologically valuable and cover different biogeographical regions (18:24).

Some people are still involved in the MAB programme since the 1970s and have continued the natural science approach. Since the world leaders have adopted the 2030 Agenda, UNESCO Biosphere Reserves can demonstrate how sustainable development can be achieved in practice at the local level. This modern and more trans-sectoral approach represents a shift in the MAB Programme (10:25). Therefore, not only the representativeness of natural areas or bioclimatic regions is needed, but also new forms of human-environment relations with a strong desire for sustainability (25:38). This reflects the wide biodiversity and structural diversity of most countries (27:47). For example, entire villages may want to achieve complete energy self-sufficiency, or urban biosphere reserves could function as model regions for sustainable development. However, it is important to maintain a certain representativeness; otherwise, the idea of biosphere reserves would be lost, and the sites would be interchangeable and arbitrary (25:39).

The world requires model regions for sustainable development. It is important not to lose track of the idea of protecting ecosystems or ecologically valuable areas, but this is only achievable if local populations have the means to survive. This is why biosphere reserves as model region for sustainable development are definitely needed, and not in short supply (18:24). The goal should be to determine how sustainable development can be achieved at the regional level and the lessons that can be learned from this at a wider scale. It would be a great success if biosphere reserves could achieve this (1:29; 31:17). New post-Seville biosphere reserves that function well generally stem from local stakeholders understanding their special status and desiring recognition for them. They are willing to invest effort and money into the UNESCO Biosphere Reserves designation and obtain international recognition for doing things well while preserving their specific identity (1:30). There is also a need to use biosphere reserves as a participatory land use planning tool that involves diverse strategies (13:26).

In addition, it would be interesting to examine agrobiodiversity in some regions. Especially in the Global South, there would be opportunities for testing agrobiodiversity initiatives, as in these countries actions focus rarely on cultural landscapes but more on natural monuments. It is mostly about turning national parks into biosphere reserves, which is not always feasible. However, for the goal to ensure sustainable management of wider natural landscapes there is significant potential in the daily cultivated communal areas, which also contain significant biodiversity (13:28).

Wildlife authorities in the Global South, for such as in East Africa, have had difficulty keeping human populations out of protected areas. Biosphere reserves are a great model in this regard because they do not completely exclude people from natural resources; with zonation, certain areas are dedicated to nature conservation and others engage in both conservation and utilization. With the paradigm shift in working with communities rather than pure conservation, it is possible to provide benefits to local communities and therefore ensure sustainability. Therefore, UNESCO Biosphere Reserves are a viable solution (9:28).

Every local initiative that aims to become a model region for sustainable development is crucial for the world and for the future. However, a challenge is to compile lessons learned from biosphere reserves and share them with other parts of the world. There should be different approaches to design sustainability challenges and thematic actions in biosphere reserves, and they should represent different natural and geographical characteristics (10:24;

26:20). One interviewed expert said the sustainability scientists have formalized the biosphere reserve concept and shown how the model can work. So far, existing biosphere reserves are not universal models. They usually consist of agricultural and peripheral areas in the countryside, not urban or industrial areas. However, at least for agricultural and remote areas, the model of biosphere reserves can be very useful and developed as sites of excellence in sustainable development and the achievement of the 2030 Agenda (16:19).

7 Synthesis and discussion

7.1 The big picture

The objective of this research relates to the ubiquitous geospatial tension between the protection of natural resources and their sustainable use. Thus, UNESCO Biosphere Reserves, a global instrument for spatial planning in mostly peripheral areas, were investigated in detail. Biosphere reserves aim to combine nature conservation and human use of natural resources based on the delineation of different zones; thus, they offer an innovative and appropriate framework for the achievement of the SDGs. The MAB Programme is the UNESCO's longest-standing intergovernmental programme to enhance the relationship between people and the environment within a scientific framework.

However, there was a significant paradigm shift in the 50-year history of UNESCO Biosphere Reserves. During the first 25 years of the programme, biosphere reserves mainly involved research on interactions between humans and the environment, with a clear focus on nature conservation. After the second World Conference on Biosphere Reserves in Seville in 1995, biosphere reserves became a spatial model for sustainable development in often peripheral natural areas. Due to this change in objectives and requirements, hundreds of biosphere reserves in the World Network had to adapt to the new framework or no longer met the criteria. As an intergovernmental UN programme, the MAB Programme faced the major challenge of how to respect cultural, social, natural and economic diversity and the national sovereignty UNESCO member states whilst ensuring a uniform framework to guarantee a certain level of quality and credibility.

Periodic reviews are the key instrument for the continuous review of compliance with the framework conditions and quality management in the World Network of Biosphere Reserves. All designated biosphere reserves must prepare and submit a periodic review to MAB governing bodies every 10 years. The MAB Secretariat describes the periodic review process as a time to take stock of progress made by the biosphere reserve, update zonation and question the objectives and means of management policies. The periodic review process also enables the discussion of weak points and improvements in the quality of biosphere reserves and their functioning as test sites for sustainable development (UNESCO, 2017b). After the significant changes to the concept after the Seville conference in 1995, such a process is particularly important (PRICE et al., 2010, p. 550).

At the third World Congress of Biosphere Reserves in Madrid in 2008, the implementation of the periodic review process was widely discussed. It became clear that biosphere reserves designated before 1995 did not meet the criteria and zonation stipulated in the revised Statutory Framework of UNESCO Biosphere Reserves (UNESCO, 2008, p. 9). After several years of discussion in the international decision-making body of UNESCO's MAB Programme and many letters from the MAB Secretariat to biosphere reserves and member states, the MAB ICC adopted the Exit Strategy to begin 'a process to ensure the continued adherence of the sites established as biosphere reserves to the objectives of their establishment and to ensure the credibility and coherence of the World Network of Biosphere Reserves' (UNESCO, 2013, p. 39). It was decided that the affected sites would be asked three more times to make the necessary adjustments and to send revised (or first-time) periodic reviews. If this did not happen, they risked being excluded from the World Network.

Some experts interviewed for this dissertation described the adoption of the Exit Strategy as overdue and necessary to clean up the network and establish a certain quality standard. Moreover, the explicitness of the threat of ejection from a UN programme was also seen as astonishing and provocative. Therefore, this thesis addresses whether current biosphere reserves in the World Network meet the quality standards of the Statutory Framework as model regions for sustainable development and assesses global quality enhancement strategies introduced by UNESCO Biosphere Reserves since 2013.

In addition, comparisons between different nature conservation designations and categories are very common, especially UNESCO labels. The latter include the strongly branded and world-renown UNESCO World Heritage Sites and the more recent UNESCO Global Geoparks, which were introduced around five years ago. Given UNESCO's tense financial situation and the renewed withdrawal of two financially stable member states, the United States and Israel, from UNESCO on 31 December 2018, there is a long-standing discussion about the financial and human resources needed to implement the three programmes and how they relate to each other. Are they compatible with each other (as in the case of Jeju Island in the Republic of Korea) or in competition with each other? This refers both to the challenge of getting more attention from society and politicians, and to funding at the international level in Paris, national governments and local implementation. Therefore, this thesis also sheds light on the interlinkages between UNESCO Biosphere Reserves, natural World Heritage Sites and Geoparks.

The 21 country-specific case studies presented in this thesis provide new insights on the great diversity of the World Network of Biosphere Reserves and different approaches to quality management and the national realization and elaboration of periodic reviews. The case studies cover all continents and thus ensure geopolitical diversity within the study sample. In addition to an in-depth examination of quality management, the implementation of periodic reviews, the effects of quality enhancement strategies and interlinkages with other UNESCO labels, interviews with national experts and decision makers revealed other aspects that are implemented differently in the world network and reflect diversity in the implementation of biosphere reserves. These include national affiliations of the MAB Programme, the legal recognition of biosphere reserves in national legislation, the naming of biosphere reserves and their governance structures.

National affiliations for the MAB Programme

An important aspect of differences in the national implementation of the MAB Programme is revealed when the allocation of responsibilities is examined, which has hardly been covered in the literature so far. Whilst Jungmeier et al. (2021) claimed that many MAB National Committees or focal points remained anchored in scientific disciplines and academia, findings from the expert interviews conducted for this thesis contradicted this statement and demonstrated that, even after the Seville Strategy and the introduction of quality enhancement measures, responsibility for the MAB Programme widely varied across different countries.

In addition to academies of science, responsibility for the programme in the case study countries predominantly fell to environmental ministries. The MAB Programme also had links to National Commissions for UNESCO and ministries of foreign affairs, as well as funding from ministries of education (see Figure 34).

In Kenya and Japan, the MAB Programme was managed by the National Commission for UNESCO, which was affiliated with the Ministry of Education (9:1). In Japan, the Ministry of Education, Culture, Sports, Science and Technology even served as the national MAB Secretariat (21:1). A In Austria, Bulgaria, China, Indonesia, the Russian Federation and Slovakia, the MAB Programme was overseen by national academies of science (ÖSTERREICHISCHES NATIONALKOMITEE MAB PROGRAMME, 2021; 15:1; 7:1; PURWANTO et al. 2020, p. 44; 17:1; 2:20). In addition, the Ministry of Environment exerted a strong influence in Bulgaria, as most biosphere reserves had previously been nature conservation areas (15:8).

Figure 34: National affiliations for the MAB Programme based on the case study countries

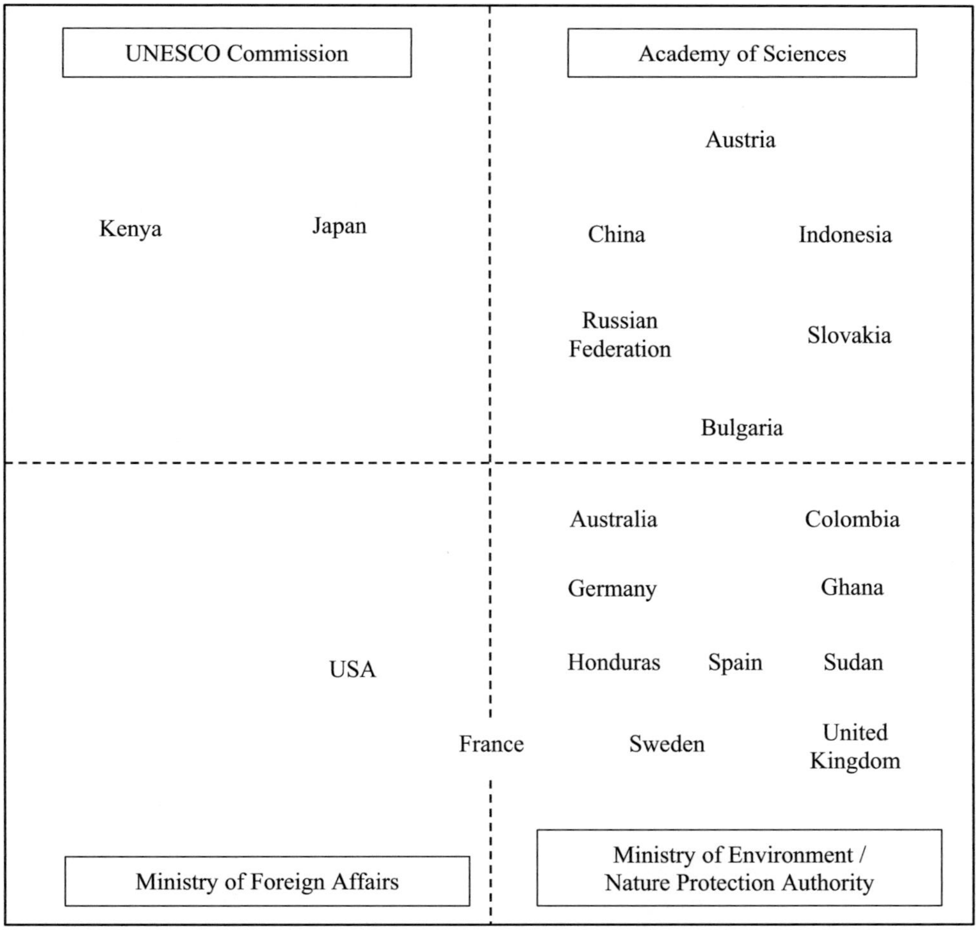

Source: Own illustration

In most countries considered in this study, responsibility for the MAB Programme and UNESCO Biosphere Reserves was linked to the ministry of environment or the nature conservation or protection authority. In Ghana, the Environmental Protection Agency hosted the MAB National Committee (3:1). In Sudan, the authority responsible for the MAB Programme was the Wildlife Conservation Administration (20:13). In Germany, the MAB National Committee was chaired by the Federal Ministry of the Environment, Nature Conservation and Nuclear Safety and managed by the Federal Agency for Nature Conservation (BfN, 2020). In Spain, responsibility laid with the National Parks Autonomous Agency, which is part of the Ministry for the Ecological Transition and the Demographic Challenge (26:3). In the United Kingdom, the Department for Environment, Food and Rural Affairs is responsible for the MAB Programme (1:4). In Colombia, the MAB Programme is managed by the Ministry of Environment with the collaboration of the Colombian National Parks Authority (23:1) and in Honduras the UNESCO Biosphere Reserves are overseen by the Institute for Forest Conservation and Development, Protected Areas and Wildlife (8:8). Two notable cases are the Republic of Korea and Sweden. In the Republic of Korea, the MAB National Committee was relocated from the National Commission for UNESCO to the Ministry of Environment in 2010; consequently, the national MAB Secretariat moved to the Korean National Park Service (MAB National Committee of the Republic of Korea, 2015, p. 2). In the same year, in Sweden, the Environmental Protection Agency took over as host for the MAB National Committee; the previous host was the Swedish Academy of Natural Sciences (4:1; 4:5). In the United States of America, the State Department is responsible for the programme and serves as the national focal point; it oversees foreign affairs, whilst the National Parks Service and the Forest Service play a strong local management role (28:1). France has a new governance structure as of 2016. Previously, the MAB Programme was affiliated with the French Commission for UNESCO, but it is now an independent association called MAB France. The chair is a scientist who was nominated by the French Ministry of Foreign Affairs, and the director of MAB France is paid by the French Agency for Biodiversity, which was also created in 2016 (24:5; 24:8).

This summary clearly shows the wide variation in the MAB Programme's affiliations in UNESCO member states. Whilst the generalization of the results is limited due to the small sample size, it would be interesting to investigate the influence of different responsibilities in ministries or academies of science on the implementation and activities of biosphere reserves in future research. This is particularly relevant given the changes implemented since the Seville conference. Whilst the focus of this work, the MAB IAC and the MAB ICC in recent years has mainly been on zoning and periodic reviews, investigating the affiliations of the MAB Programme in different countries should also be considered as a point of discussion.

Legal recognition of biosphere reserves in national legislation

Another frequent topic of discussion at the international level was the legal recognition of biosphere reserves in national legislation, which can also be seen as a dispute over direction and again reflects diversity of implementation in the World Network of Biosphere Reserves. Whilst the core zone of all biosphere reserves must be a state-protected nature conservation area, the legal framework for buffer and transition zones is not precisely defined in the

Statutory Framework. PRICE (2017) argued that, to meet the objectives of the 10-year Lima Action Plan, all biosphere reserves require an effective and participatory governance structure with support from all governance levels. This support is most evident in countries where biosphere reserves are recognized in national legislation (p. 38). In addition, ISHWARAN et al. (2008) indicated that 80% of designated biosphere reserves were not integrated into national legislation on protected areas; thus, biosphere reserve managers usually did not have responsibility beyond the core areas. However, providing relevant authorities with the governmental mandate and necessary resources to address stakeholder interests in biosphere reserves would be key to the success of the World Network of Biosphere Reserves (pp. 123-125). In a comparative analysis of legal frameworks for biosphere reserves in Ukraine and Sweden, ELBAKIDZE et al. (2013) concluded that there is a need for differentiated and adapted solutions on the ground in different societal contexts, given the goal of transforming biosphere reserves into learning sites for sustainable development. Whilst the core functions of biosphere reserves were supported by legal documentation in both countries, legal support for the promotion of sustainable development remained unclear (p. 185).

The findings from this study align with those of ISHWARAN et al. (2008), as 15 out of 21 case study countries did not include biosphere reserves in national legal frameworks (see Table 12). In three countries, biosphere reserves were mentioned in law but not as a separate category of nature conservation area; usually, they were mentioned only vaguely, without concrete management requirements or responsibilities. Biosphere reserves were legally recognized in national legislation in only three countries: France, Germany and Spain. Whilst these countries treated the entire biosphere reserve as a protected area, there were enough opportunities to implement biosphere reserves under existing legislation (13:21). However, several interviewed experts mentioned that the classification of an entire biosphere reserve as a protected area could also be a disadvantage when communicating with local populations because many people have already had negative experiences with other protected area designations and were not familiar with the concept of biosphere reserves. In such cases, they usually did not want a new protected area (27:9).

In this regard, two special cases are China and Germany. In both countries, national biosphere reserves existed alongside UNESCO Biosphere Reserves. Since biosphere reserves in Germany are designated as a separate protected area category in the Federal Act for the Protection of Nature and in federal states, they could be designated independently from UNESCO recognition. Although the MAB National Committee aims to ensure that all biosphere reserves designated by federal states are also designated by UNESCO, only 16 out of 18 German biosphere reserves are currently recognized by UNESCO and part of the World Network (5:5, 27:8). In China, a national network of biosphere reserves was introduced in 1993, which has grown to 177 sites. These exist alongside the 34 UNESCO Biosphere Reserves. As in Germany, they are designated under the national system prior to UNESCO nomination. However, competition between biosphere reserves is intense, and only sites that have already been active in the national network for several years can be selected for UNESCO nomination and participation in the World Network (7:17).

Discussions about legal recognition for biosphere reserves in national legislation also revealed a clash of values in the MAB network, as different disciplines variously view biosphere reserves as conventional protected areas, such as national parks, game reserves and forest reserves, or a recognition of local efforts to engage in the sustainable management of natural resources. In countries where biosphere reserves are managed by the Ministry of the Environment or protected area authorities, biosphere reserves were treated more as conventional nature conservation areas. In other countries, the 'biosphere reserves' designation is seen as a great opportunity for best practices in sustainable use of natural resources in harmony with nature conservation. The main reason was the non-designation as a nature conservation category and thereby all its negative perceptions amongst local populations in and around existing protected areas.

Usage of the term 'biosphere reserve'

Another discussion that has been ongoing since the inception of the MAB Programme is the use of the word 'reserve' to describe biosphere reserves. Whilst MAB governing bodies allow member states to use their own terminology for biosphere reserves, STOLL-KLEEMANN and O'RIORDAN (2018) argued that the term has rather negative connotations and conveys an area that people are excluded from. They called for the term 'biosphere reserve' to be replaced by 'biosphere landscape' (pp. 350, 352). The experts interviewed for this thesis demonstrated the diffuseness of the term 'reserve' in the World Network. Only half of the 21 case study countries considered in this thesis used the full term 'biosphere reserve' (see Table 13). In Australia and Bulgaria, the term 'reserve' was linked to strict protected areas without public access under state legislation and was thus not used. In many other Western countries, the term 'reserve' evoked reservations amongst local populations; as a result, biosphere reserves were simply called 'biospheres' in Australia, France, the United Kingdom and the United States. In Estonia, Sweden and parts of Germany, they are called 'biosphere areas'; in Austria and Bulgaria, they are called 'biosphere parks' in accordance with the protected area category 'national park'. The most significant naming deviation was in Japan, where the MAB National Committee changed renamed biosphere reserves to 'UNESCO Eco Parks' in 2010. In African, Arab State and Latin American case study countries, by contrast, there were fewer problems with or deviations from the term 'biosphere reserves'. In future studies, it would be interesting to investigate why the word 'reserve' encounters resistance in parts of the world where protected area categories such as game reserves or national reserves have also existed for some time.

Since most experts interviewed for this thesis stated that one of the greatest current challenges in the MAB Programme is a lack of awareness and recognition of the biosphere reserve concept, it is important for the MAB Programme to enforce uniform terminology for biosphere reserves worldwide. Competition with other protected area categories and international designations such as UNESCO World Heritage Sites and Geoparks make it more difficult for biosphere reserves to be globally recognized; this should not be exacerbated by the fact that nomenclature is different in various member states. Thus, STOLL-KLEEMANN and O'RIORDAN's proposal to rename biosphere reserves to 'biosphere landscapes' or an interviewee's proposal to rename them (21:16) 'Man and the Biosphere Sites' or 'MAB Sites' should be seriously considered by the MAB ICC in a timely manner. Although it might be risky to change the name after 50 years, it could also be a tremendous opportunity for biosphere reserves to play a leading role in the post-2020 agenda as model regions for sustainable development, as they would have a fresh start and shed misconceptions. If the MAB

ICC decides not to change the name, it is important to standardize variations in terminology to create a unified brand and promote the visibility of the World Network of Biosphere Reserves.

Governance structure of biosphere reserves

There is no single model when it comes to governance structures for biosphere reserves. According to the interviewed experts, some biosphere reserves are managed entirely as nature conservation areas, whilst others are community-run and have independent management and governance structures that include all relevant stakeholders. This mix of different governance structures also exists within the states, where in most cases there is rather a combination of the different systems. However, the findings also show that first-generation biosphere reserves in particular are still managed as classic nature conservation areas.

In Germany, biosphere reserves are included in the Nature Conservation Act as a separate protected area category and managed as protected areas. This is the only case in the present global analysis of governance structures. In France, most biosphere reserves are integrated into national nature conservation law as regional nature parks and managed as protected areas. Only two biosphere reserves are community-run. In Ghana, Kenya and Sudan, there is a mix of management structures; most biosphere reserves are protected areas, usually national parks, and thus managed by the relevant protected area authorities. For example, at the Mount Kenya Lewa Biosphere Reserve, there is no dedicated manager for the entire biosphere reserve; the core zone is managed as a national park by the Kenya Wildlife Service, and the buffer and transition zones are overseen by the Kenya Forest Service. Coordination between the two institutions is conducted by the MAB National Committee. In addition, one biosphere reserve in each of the three abovementioned countries is entirely managed by local communities (see Chapter 5).

In Sweden and the United Kingdom, some NGOs or foundations are responsible for the entire management of biosphere reserves and act as umbrella organizations for related activities. Therefore, the staff is employed by the NGOs; governmental agencies, charities or other organizations can provide funding. Another governance system relies on support from local governments. For example, a biosphere reserve in Sweden is part of a municipality administration; it has an independent office funded by the municipality's budget (10:4; 1:11).

Biosphere reserves in China and the Republic of Korea are overseen by a management committee and financed by local governments. In China, management committees consist of bureaus for the protected area, local governments and local communities and stakeholders (7:10; 7:12). Management committees in the Republic of Korea have a similar structure and include local governmental institutions, local communities, representatives from the MAB National Committee, local NGOs and initiatives and academics from universities (12:7).

In Spain, the country with the most UNESCO Biosphere Reserves in the world, there is also a mix of governance structures. In Andalusia, Extremadura, Asturias and Madrid, autonomous authorities manage biosphere reserves as protected areas, mostly nature parks or national parks. Other sites are managed by local administrations, as in the Canary Islands, or provincial councils, as in Catalonia, Barcelona and Galicia. Finally, another option for state-run biosphere reserves in Spain is management by *mancomunidades*. Moreover, one biosphere reserve in Leon is managed by an NGO, which is an association of cattle breeders (26:15).

In this thesis, different approaches to national affiliations of the MAB Programme, the legal recognition of biosphere reserves in national legislation, usage of the term 'biosphere reserves' and the governance structures of biosphere reserves demonstrate the diversity of the World Network of Biosphere Reserves at a superordinate structural level. This diversity is crucial for understanding the impact of recent quality enhancement strategies on different UNESCO member states, which is presented and discussed in the following sections.

7.2 National implementation of quality management requirements and the periodic review process

The Statutory Framework of UNESCO Biosphere Reserves (1996) is still considered the basis and main source of guidance for the implementation of biosphere reserves by member states. Whilst the framework's 10 articles establish basic aspects such as definition, functions and the participation of member states and the MAB Secretariat, they do not contain concrete guidelines for quality management in the World Network of Biosphere Reserves. Quality management is only mentioned in Article 9, which states that 'the status of each biosphere reserves should be subject to a periodic review every 10 years, based on a report prepared by the concerned authority (…) and forwarded to the secretariat by the State concerned' (UNESCO 1996: 18). Whilst the following points of the Article 9 only concern evaluation within the MAB governance system, they do not specify the concrete implementation of 'preparation by the concerned authority'. The template for the periodic reviews also does not further indicate how they should or could be conducted within member states.

Notably, MAB National Committees are not explained in detail in the Statutory Framework but only mentioned in Article 5 on the designation procedure, which specifies that 'states, through MAB National Committees where appropriate' (UNESCO, 1996, p. 17), should forward the nomination of potential sites to the MAB Secretariat. In a 1997 MAB ICC handout, guidelines for the establishment of MAB National Committees were presented; the handout emphasized that the decision to create a MAB National Committee and the means for its establishment depend on the internal organization of each state; thus, their composition varies from country to country. It was also stipulated that, in the absence of a MAB National Committee, the MAB focal point would take over its functions. The main role of the MAB National Committee is to be responsible for national activities under the MAB Programme, particularly biosphere reserves. It serves as an intermediary that liaises between the UNESCO National Commissions, institutions and ministries involved in the MAB Programme and UNESCO (UNESCO, 1997, p. 1). Although national quality management is not defined in the Statutory Framework, the MAB National Committees play a crucial role in the implementation. To understand the implementation of quality management and periodic reviews and thus the impact of the Exit Strategy in different countries, the first research question of this thesis is as follows: 'What are differences in quality management and the periodic review process of biosphere reserves between member states?'

Based on expert interviews conducted for this thesis, the analysis showed that two out of 21 case study countries (both in Latin America) still have not established a MAB National Committee. In Colombia, there has been a long-standing discussion about whether to create an independent MAB National Committee or use existing structures within the government. So far, the Ministry of Environment is responsible for providing all relevant information for the periodic reviews and conducting diplomatic negotiations in the MAB ICC. However, since this is very challenging, Colombia's current priority is to establish a MAB National Committee (23:4). Honduras has the same goal. Although a memorandum of understanding between the National Commission for UNESCO, the Department of Natural Resources, Mines and the Environment and the Ministry of Foreign Affairs was signed in 2011 or 2012 to establish a MAB National Committee, Honduras still has not established one (8:4; 8:12). All other countries surveyed on the other continents had established a MAB National Committee.

The composition of the MAB National Committees is, in all investigated countries that have established one, in line with the guidelines, which advised that

> 'in order to take account of the interests of the scientific community and the administrative authorities, the MAB National Committee should be composed of representatives of the main scientific research centres, and of the universities and ministries concerned, be interdisciplinary (...) [and] the authority in charge of each biosphere reserve should also be represented on the MAB National Committee'. (UNESCO, 1997, p. 2)

In Sweden, the committee only includes five members from the Environmental Protection Agency, the Forestry Agency and the Stockholm Resilience Centre; the National Commission for UNESCO and the Oceans, Marine and Water Authority also attend meetings (10:9). Otherwise, most MAB National Committees consist of 20 to 30 members; for example, there are 20 members in Germany, 25 in the Republic of Korea and 28 in Ghana. In Spain, there are 34 members. In general, meetings take place twice a year. In Austria, Germany and Sweden, one meeting per year usually takes place at one of the biosphere reserves.

In addition to MAB National Committees, Kenya, Spain and Sweden have established other committees or councils. In Kenya, an expert committee was created to ensure the implementation of activities for nature conservation, raise awareness and support the development of designated biosphere reserves. In Spain, there are two advisory bodies, a scientific council and a management council that complement the work of the MAB National Committee. The management council includes one technical representative from each biosphere reserve, whilst the scientific council only has two representatives who prepare the periodic reports. When a proposal for new biosphere reserve is submitted, it is first sent to the scientific council. The board is independent and analyses the proposals. In addition, it monitors the biosphere reserves through an indicator system. The latter was approved in 2010 at a joint meeting between managers and scientists. The analysis consists of eight indicators and is done annually. Results are sent to the MAB National Committee to provide them an overview of the annual average value of the status of implementation of actions and requirements of the MAB Programme for each site (26:5). The indicator system was also shared with Portugal, Morocco and IberoMAB for any members states that want to use it (26:7). In Sweden, the Biosphere Reserve Council was created in addition to the MAB National

Committee, which consists of all coordinators at biosphere reserves and the national coordinator. The role of the council is to communicate the needs of the biosphere reserves to the MAB National Committee, which acts as an advisory body to the MAB National Committee for new biosphere reserve nominations (4:10; 10:10).

In most of the case study countries, periodic reviews were conducted in cooperation with biosphere reserve managers and the MAB National Committee. In Kenya, for example, the MAB National Committee had the lead on writing the periodic review report (9:8), whilst the periodic reviews in France, the Republic of Korea and Sudan were conducted by biosphere reserve teams with support from the MAB National Committee (24:10; 12:11; 20:11). Notably, the evaluation of biosphere reserves for the periodic review process in Japan was conducted by voluntary research groups under the Japanese MAB National Committee (21:3). In Estonia, by contrast, periodic reviews in 2005 and 2015 were conducted by subcontracted consultants (16:1). In France, students are often involved in the periodic review process in addition to local biosphere reserves managers and the MAB National Committee. At one university in Toulouse, there is a master's programme that focuses on the MAB Programme; thus, a group of students often helps organize discussions with local people during the periodic review or the nomination process for new sites (24:10).

Two special cases are China and Germany. In Germany, over the past 20 years, the MAB National Committee has customarily visited biosphere reserves the year before the 10-year periodic reviews to compile its own evaluation report, in addition to the preparation of periodic review reports by the sites themselves. In addition, the MAB National Committee writes a letter to the MAB Secretariat with its own recommendations; this accompanies the periodic reviews (25:36). The evaluation and recommendations of the German MAB National Committee provide an external view on the work of German biosphere reserves, which is taken quite seriously (27:24). In China, an internal, national periodic review process begins several months before the formal review. A small group of experts and staff from the China MAB Secretariat travels to the biosphere reserves to assess the situation, give managers feedback on their performance and propose advice and improvements before the formal periodic review (7:6). For the formal periodic review, a larger group of experts and staff from the MAB China Secretariat travels to biosphere reserves and holds several meetings with local people and the staff of the protected areas. Afterwards, the experts hold an internal meeting to develop a joint review, which is then presented to the biosphere reserve staff, the leadership of local governments and local villagers in a conference to discuss the main achievements of the last 10 years and the direction that future activities should take. The summary this process is then translated into English and submitted to UNESCO (7:5; 7:21).

In addition to the diverse national affiliations of the MAB Programme, legal recognition in national legislation, usage of the term 'biosphere reserves' and the governance structures of biosphere reserves, the findings show that there was diversity in the implementation of quality management measures and performance in the periodic reviews within member states. These results should be considered when examining the Exit Strategy's effects on different member states, as national implementation and enforcement likely influence the impact of quality enhancement strategies. Despite the Statutory Framework, the MAB Programme and the World Network of Biosphere Reserves do not form a unified system.

7.3 Effects of recent quality enhancement strategies on MAB Programme

An analysis of the MAB Secretariat in 2012 revealed that 287 biosphere reserves had conducted a periodic review; however, 61 did not meet the criteria of the Statutory Framework for Biosphere Reserves. In addition, 112 designated biosphere reserves had never performed a periodic review. The MAB Secretariat sent letters to request that these biosphere reserves conduct a periodic review and to follow up on recommendation issued by the MAB ICC. Whilst 145 biosphere reserves answered in time, 131 biosphere reserves did not respond to the letters (UNESCO, 2013, pp. 37-38). Because this situation was no longer tenable, the MAB ICC adopted the Exit Strategy in 2013 as 'a process to ensure the continued adherence of the sites established as biosphere reserves to the objectives of their establishment and to ensure the credibility and coherence of the World Network of Biosphere Reserves' (UNESCO, 2013, pp. 38-39). The affected biosphere reserves and their states were granted a last chance to respond to the warning letters. In case there was no reply, the Exit Strategy recommended that the MAB ICC remove the sites from the World Network of Biosphere Reserves (UNESCO, 2013, p. 39).

The threat of exclusion was a unique approach for an intergovernmental programme of the UN and comprised a wake-up call for many member states, with major implications for the national implementation of the MAB Programme and biosphere reserves. Therefore, the second research question in this thesis was as follows: 'What effects did the Exit Strategy and later the Process of Excellence and Enhancement have on the MAB Programme in general and the 76 directly affected member states and their 266 biosphere reserves in particular?'

Since the beginning of the MAB Programme in the 1970s, 13 UNESCO member states have voluntarily withdrawn 59 biosphere reserves from the World Network of Biosphere Reserves; 46 of these sites were withdrawn from 2013 to 2020, after the Exit Strategy was implemented. Most withdrawals were made by the United States (19 sites), followed by Australia (10 sites), Bulgaria (seven sites), Austria (four sites), and the Central African Republic, the Democratic Republic of Congo, Denmark, Mexico, the United Kingdom and the Netherlands (one site each; UNESCO, 2020c; UNESCO, 2020a, p. 2; UNESCO, 2020b, pp. 2-4). In this thesis, a special focus was placed on some of these countries, with expert interviews and detailed case studies of the United States, Australia, Bulgaria, Austria and the United Kingdom. These case studies are examined and presented in more detail in Chapter 5. In combination with the other countries included in this study, which also were impacted by the Exit Strategy but with different consequences than the voluntary withdrawal of sites, these results provide new insights on the effects of the Exit Strategy. These are important for the future direction of the MAB Programme and the influence of international nature conservation designations on the national and regional implementation of nature conservation measures.

In Australia, the United States of America, Austria and Bulgaria, the Exit Strategy reinforced the national process of addressing first-generation biosphere reserves that did not align with the concept of biosphere reserves as defined in the Seville Strategy (for detailed country-specific descriptions, see Sub-sections 5.3.1, 5.4.10, 5.4.1 and 5.4.2). Between 1976 and 1982, 12 biosphere reserves were designated in Australia; like many first-generation biosphere reserves, they focused on nature conservation and, in some cases, had no human settlements. After their designation, not much happened. In 2002 and respectively in 2011, the Southwest National Park Biosphere Reserve in Tasmania and the sub-Antarctic Macquarie Island Biosphere Reserve were withdrawn from the World Network of Biosphere Reserves, as they were not compatible with the Seville Strategy. With the adoption of the Exit Strategy, the focus turned to the future of Australia's remaining first-generation biosphere reserves. Apart from the Fitzgerald and Riverland Biosphere Reserves, the remaining eight biosphere reserves did not conform to the Statutory Framework. Thus, the Yathong, Prince Regent River, Hattah-Kulkyne and Murray Kulkyne, Wilsons Promontory and Barkindji Biosphere Reserves were withdrawn in 2018, and the Uluru, Kosciuszko, Unnamed and Croajingolong Biosphere Reserves were withdrawn in 2020 (UNESCO, 2020c).

The Riverland Biosphere Reserve was renamed twice (once in 1994 and once in 2004) and tried to implement the concept of a model region for sustainable development for several years. However, the complexity of management bodies, which involved the federal government and local communities, and increased disinterest from the government meant that the second periodic review was overdue for several years. Ultimately, the biosphere reserve was withdrawn from the World Network in 2020 (BRIDGEWATER, 2020, pp. 233-234). The only positive example of first-generation biosphere reserves is the Fitzgerald Biosphere Reserve in Southwest Australia. The adjacent community was a very active cultural site that became interested in extending the first-generation biosphere reserve to rejuvenate farmland through innovative conservation ideas. With strong participation from the community and the national park administration, the biosphere reserve was extended with new zonation and successfully re-nominated in 2017 (14:14). The Biosphere Reserve is complemented by three community-run initiatives in Australia, which together comprise Australia's four post-Seville biosphere reserves.

Biosphere reserves in the United States of America have been affected by historically difficult political relations between the national government and UNESCO and general distrust of the UN amongst many rural private landowners, who feared attempts to interfere with their private property. In 1976, the United States of America designated 27 biosphere reserves, with 20 additional sites until 1991. At the time, the United States of America had the most designated biosphere reserves in the world. Since 1991, however, it has not designated any new biosphere reserves. When the Exit Strategy was adopted in 2013, all 47 of its biosphere reserves were first generation and thus affected by the strategy. Biosphere reserves in the United States of America mainly serve as strictly protected areas for conservation and benchmarks for monitoring ecological change. They were all managed by the National Park Service, the Forest Service or the Agricultural Research Service. This system functioned well until first the Reagan administration withdrew membership and support from UNESCO in 1984; the agencies, especially the National Park Service and the Forest Service, were under pressure from the U.S. Congress in 1996 for their involvement in the MAB Programme, which resulted in their withdrawal of financial and human resources for biosphere reserves. Revitalization attempts by Presidents George W. Bush and Barack Obama only aroused the interest of Americans in the short term and did not strengthen the MAB Programme at a national level.

In 2013, all 47 biosphere reserves in the United States of America did not submit a periodic review. Historically, zonation has been an issue in the United States of America due to the concentric structure of biosphere reserves, which include core, buffer and transition zones. Many people took issue with an international organization having any input on privately owned property. When the Rocky Mountain Biosphere Reserve conducted a periodic review in 2016, they proposed slightly different terminology to UNESCO that would be more accepted in the United States of America. The core zone remained the same, but the buffer zone was renamed an 'area of transition' or 'joint-management area'. Although it remained private property, it featured a conservation easement that protected the area around the core zone. The transition zone was called an 'area of cooperation' to assure local populations that cooperative agreements were strictly voluntary at the local, state and national levels (28:1). The following year, several other biosphere reserves in the United States of America used this terminology and passed the periodic review process. However, 19 biosphere reserves that had been inactive for a long time and had no grass roots to further continue stated that the post-Seville model of biosphere reserves was no longer mainly a science programme and voluntarily withdrew their nominations from the World Network between 2017 and 2019 (28:2).

Although many biosphere reserves were withdrawn due to the Exit Strategy, some local groups wanted to continue the work of the biosphere reserves. This was particularly noticeable in sites that had not been previously managed by the National Park Service or the federal government, such as the Champlain-Adirondak Biosphere Reserve in upstate New York, near the Canadian border. Local stakeholders tried to implement joint activities with the adjacent Canadian biosphere reserve. In addition, two biosphere reserves on the Mexican border were interested in working across the border (28:3; 28:12). In Hawaii, a biosphere reserve is looking for ways to promote the local culture of growing trees that are used in traditional canoes but were almost completely logged (28:20).

Austria designated its first four biosphere reserves in 1977 and another four post-Seville sites after 2000. The Exit Strategy was an issue at a very early stage, as it had been clear for some time that there were two different levels of quality in Austria: modern post-Seville biosphere reserves and first-generation sites, which focused on basic research and did not have a buffer zone, inhabitants or a strategy for sustainable development. The transformation of the first-generation biosphere reserves was not successful due to a lack of political will. For example, the Neusiedler See Biosphere Reserve consisted only of the actual lake; there was great fear that the enlargement of the biosphere reserve would lead to many more restrictions for local communities. For the government, the labels 'national park' and 'World Heritage Site' were sufficient, and they did not see added value in the site's designation as a biosphere reserve. Thus, all four first-generation biosphere reserves were withdrawn from the World Network in 2014 and 2016.

Another country that was one of the first to join the MAB Programme was Bulgaria. In 1977, Bulgaria designated 17 biosphere reserves, making it one of the countries with the most sites in the World Network at the time (15:5). However, Bulgaria was not very active in managing its first-generation biosphere reserves or adapting them to the new paradigm after the Seville conference in 1996. It did not dedicate any policy to transform and adjust the biosphere reserves to the new criteria, zoning system and functions. Fifteen biosphere reserves were strict nature reserves without a transition area (15:6). One of the reasons was a focus on the development of Natura 2000 sites, which was a condition for Bulgaria to join the EU in 2007. As a result, there was not much time and commitment for meeting the requirements of the MAB Programme. Another important issue was political changes after the communist period in 1990. State-owned lands were returned to their previous owners, who did not want any more restrictions on their newly returned properties (15:7; 30:11). After Bulgaria received the first warning letters from the MAB Secretariat after the adoption of the Exit Strategy, the Ministry of Environment and Water, the MAB National Committee and the Bulgarian Biodiversity Foundation worked together to renew the biosphere reserve network in Bulgaria. Funds were provided by the ministry, but there was only enough to focus on some of the country's 16 biosphere reserves (one site had already been withdrawn in 2002). After many meetings with mayors, local communities and municipalities, only four sites were selected: Central Balkan, Srébarna, Tcheryenata sténa and Ouzounboudjak. All of them were successfully transformed into post-Seville biosphere reserves (15:9; 15:10). The newly created Central Balkan Biosphere Reserve encompassed four biosphere reserves that had been previously designated in 1977 (30:5). All other first-generation biosphere reserves were withdrawn from the World Network in 2018, 2020 and 2021. For Bulgaria, the adoption and implementation of the Exit Strategy was a positive initiative, as its firm deadline was a wake-up call for the ministry and the adaptation of the biosphere reserves became a new priority (15:27).

The four countries with the most voluntary withdrawals of biosphere reserves – the United States, Australia, Austria and Bulgaria – demonstrate how the Exit Strategy contributed to the streamlining and quality enhancement of the World Network after the paradigm shift. In these countries, only post-Seville biosphere reserves remain. These nations designated biosphere reserves early on in the MAB Programme and the voluntarily withdrawn first-generation biosphere reserves were mainly strict nature conservation areas without any human settlements. Although all member states were aware of the inconsistency with the Statutory Framework and the new orientation of the MAB Programme, political pressure to act only arose with the implementation of the Exit Strategy and the MAB Secretariat's threat to forcibly remove non-conforming biosphere reserves from the World Network.

The effects of the Exit Strategy were not limited to countries with voluntary withdrawals. Other member states also had problems with first-generation biosphere reserves that were affected by the Exit Strategy. In Ghana, the pre-Seville Bia Biosphere Reserve was a typical national park. It was successfully rezoned, and a new nomination was submitted (3:1). In Kenya, Malindi-Watamu Biosphere Reserve had to add a new terrestrial core zone to achieve the required zonation but lacked the necessary funding to tackle the rezonation and submission to the MAB Secretariat (29:1). In Slovakia, the main problem was also a lack of funding to hire someone to assist in the adaptation of first-generation biosphere reserves to the post-Seville criteria (2:4). Kiunga Biosphere Reserve in Kenya and Radom Biosphere Reserve in Sudan were also affected by the Exit Strategy; however, both are part of an armed conflict, and the decision was made to keep both sites in the network until it is resolved (29:4; 20:16). In Indonesia, first-generation biosphere reserves were identical with the area of existing nature conservation areas (mostly national parks) and thus also had to undergo rezonation. In addition, in the Republic of Korea, the only first-generation biosphere reserve, Mount Sorak, was extended with a larger transition area (12:16), as was the case in Spain at the Las Cuentas Altas Manzanares Biosphere Reserve (26:17). In China, a special task force was appointed to perform the necessary adjustments and re-nominations for four Chinese biosphere reserves affected by the Exit Strategy. For example, the Dinghushan Biosphere Reserve had issues with zonation and community involvement (7:26).

In Japan, all four first-generation biosphere reserves established in 1980 were also affected by the Exit Strategy. These sites were chosen as part of a top-down approach and were mostly paper parks with barely any actions or activities; thus, the Exit Strategy was a good opportunity to revitalize Japanese UNESCO Biosphere Reserves. The Japanese MAB Secretariat urged local governments from the four sites to decide whether to keep their designation under the new criteria. All four decided to maintain their nomination and established new management bodies called Biosphere Reserve Councils, which are led by local governments and involve diverse local stakeholders (21:5). A similar situation was also reported during the expert interviews in France, where the Exit Strategy was also perceived as a good quality enhancement measure. At the two affected biosphere reserves, the Exit Strategy was an important signal and pressure point to avoid losing the label. For one biosphere reserve, zoning was revised, and the nomination was successfully resubmitted; at the second site, the missing management plan was developed (24:14).

These results build on existing evidence from COETZER et al. (2014) and MATAR and ANTHONY (2017), who indicated that most of the MAB IAC's recommendations on periodic reviews were linked to incomplete or inefficient zonation, insufficient community involvement and inadequate management plans for the biosphere reserves. Moreover, the greatest challenges were a lack of technical and financial capacity.

Moreover, the expert interviews have other dimensions of the Exit Strategy. Firstly, a shift in the membership of the MAB ICC over time was significant for the successful implementation of quality enhancement strategies. In the past, older men from the natural sciences were the main participants at MAB ICC meetings; now, a younger and more diverse generation from countries that have successfully implemented post-Seville biosphere reserves is observable (1:16). Secondly, it remains unknown who actually received the warning letters from the MAB Secretariat. It is unclear whether they were forwarded to the managers of biosphere reserves or stalled at the MAB focal points and member states' permanent representation in Paris (25:32; 27:38). This also demonstrates that diplomatic, hierarchical and bureaucratic communication within UNESCO and the MAB Programme could be improved to foster an active network of model regions for sustainable development around the world. It would also explain delays and failures to implement the paradigm shift after the Seville conference even before the introduction of the Exit Strategy.

Third, another discussion point regarding the implementation of the Exit Strategy is the political dimension, particularly what would actually happen to biosphere reserves affected by the Exit Strategy after the deadline and the procedure for the MAB ICC to issue withdrawals of designation. Some member states demanded that non-functioning biosphere reserves be removed from the World Network, whilst others asked for additional support to be given to affected sites to help them conform to the Statutory Framework (19:10.2). Because UNESCO is a member state organization and the MAB ICC is an intergovernmental body, the expulsion of biosphere reserves would be politically and diplomatically challenging. Even if countries do not sit on the MAB ICC, they are likely to have good relations with other countries that may vote for them. Thus, the question remains of how to pragmatically address involuntary withdrawals to ensure that the Exit Strategy has real consequences (25:33; 13:1).

A discussion about direction became heated during the MAB ICC meeting in 2017. There had been a proposal in the MAB Council from European countries to exclude 85 biosphere reserves from the World Network and withdraw their designations. Under the leadership of the Latin American group and a few African countries, it was voted down. The proposal's detractors argued that the biosphere reserves should stay in the network and that they should continue working on quality enhancement strategies to improve the sites or let them voluntarily withdraw from the network (19:10.2; 18:21; 7:8; 8:6; 23:2). Compromises were negotiated under negotiations of the French chair of the MAB ICC and Council. The decision was made to exclude the biosphere reserves from the World Network, but this would only come into force two years later, which would provide affected biosphere reserves with the opportunity to avoid losing their designation – not by submitting additional documents but a new application (25:33; 18:21). The ambassadors from the affected countries made it clear that the issue of the Exit Strategy and the then renamed 'Process of Excellence and Enhancement' was recognized and that it has been put on the back burner in the previous years. However, it was also important to understand that the process was not about excluding and punishing biosphere reserves but rather to permanently strengthen the other 90% of the World Network of Biosphere Reserves (27:36).

According to the interviewed experts, the greatest criticism about the implementation of the Exit Strategy was its name. Many felt that the word 'exit' had negative connotations. Any strategy should aim to improve the situation and achieve positive outcomes, not an exit. Thus, renaming the strategy to the Process of Excellence and Enhancement was important (2:9; 2:10; 1:18; 14:25; 12:17; 8:17; 29:4; 16:14; 10:18).

The analysis showed that the challenges in implementing the Statutory Framework, rezoning and the preparation of periodic reviews could not be attributed to individual countries; even within member states, there were significant differences in the implementation of biosphere reserves during the 50-year history of the MAB Programme. The United States of America, Australia, Austria and Bulgaria exemplify this.

All interviewed experts agreed that the quality enhancement strategies were overdue and fundamental to improving the credibility of the MAB Programme. Although there is still an imbalance in quality level between biosphere reserves after the Exit Strategy, it was the best strategy for adapting old, non-functioning biosphere reserves or encouraging member states to voluntarily withdraw them from the network. When comparing sites, diversity and flexibility in the general implementation of the biosphere reserve concept should be considered, as shown in Section 7.1; these factors also affect their comparability. In addition, the results showed that, due to the threat of the Exit Strategy, many member states and governments renewed their involvement in the MAB Programme, which led to a rejuvenation of the programme and network in many countries.

7.4 Improvement of evaluation process

With the Exit Strategy and the Process of Excellence and Enhancement, the MAB Programme aimed to ensure that all biosphere reserves in the World Network meet the minimum requirements of the Seville Strategy, particularly with regard to the three zones and new functions, as well as compliance with the 10-year periodic review cycle. However, the periodic review process (and thus the evaluation capacity of the MAB Programme itself) has been criticized for its length and the great effort and costs required of biosphere reserves. Therefore, this thesis also addressed the following research question: 'What kind of new ways and methods could be used to improve the evaluation process?'

The 10-year reporting cycle for periodic reviews has long been discussed; it has been criticized for being too long to effectively monitor changes and the achievement of objectives at biosphere reserves. However, five years may be too short to implement MAB ICC recommendations, such as zonation changes (PRICE et al., 2010, p. 555). After serious discussions within the MAB IAC, the idea of reducing the timeframe from 10 to five years was largely abandoned due to its limited capacity to review twice the number of periodic reviews (MATAR & ANTHONY 2018, pp. 13-14).

However, 73% of experts surveyed[2] in this thesis were in favour of shortening the reporting cycle to, for example, five years, whilst the remaining 27% wanted to keep the reporting cycle at 10 years. Advocates of the 10-year cycle argued that biosphere reserves represent a long-term commitment and entail collaboration processes that take significant time; thus, it may be a while before effects can be seen (14:21; 31:9; 24:10; 10:14). Other experts, however, said that much can happen in 10 years. Biosphere reserves are a living concept; there may be changes in the size of areas, the number of inhabitants or the management of the sites by local, regional or national authorities. These developments must be communicated to UNESCO to verify whether the biosphere reserve still complies with the criteria (19:7; 31:9). Because pressure to protect natural areas is increasing in many countries, there is a need to verify whether biosphere reserves are at risk to allow issues to be addressed as soon as possible. Waiting 10 years may be too long (29:3).

One suggestion was to introduce a five-year supplementary review with a simpler questionnaire to provide a snapshot of the status of each biosphere reserve, with basic facts about the health of the ecosystem and recent changes (29:7; 5:7). This could be a supporting process for the more detailed 10-year periodic review (31:9). In Japan and the Republic of Korea, the MAB National Committees requested all biosphere reserves to provide a short annual progress report (12:13). Biosphere reserve managers in Spain have also been collecting data on eight indicators annually since 2010 to inform the MAB National Committee about the implementation status of the MAB Programme at each biosphere reserve (26:5). Furthermore, some of the interviewed experts suggested that the periodic review template itself should be revised to facilitate periodic reviews in the future. Currently, it is quite long, and it takes significant time and work to collect and fill in all the required information. A simpler version of the questionnaire with rubrics that people could assign scores to might facilitate more regular updates (1:23; 3:7; 13:13).

2 Only experts who have explicitly commented on this topic were considered.

Another point of discussion was the self-evaluation bias of periodic reviews, as reports are compiled member states and not an external figure from IUCN, as with the World Heritage Sites. Whilst there was significant criticism about the self-evaluations, their quality ultimately depends on implementation in different member states. Some countries take the process quite seriously (24:12). One suggestion for improvement was to introduce a voluntary peer review system in which another biosphere reserve from the same country or another country in the region would also examine the report. Biosphere reserves would still perform their own review but also get another perspective. This could benefit both biosphere reserves and increase twinning arrangements in the network. A voluntary peer review could be seen as an added value in the submission of periodic reviews to the MAB IAC and ICC (10:12; 10:13; 21:11). Another option is to have an external evaluation team examine each biosphere reserve in the World Network. This approach would enable comparability and be more objective, but it is very expensive and not particularly realistic to implement given UNESCO's current financial situation (26:17).

Another idea is to involve regional MAB networks in periodic reviews. The periodic reviews could go first to a regional level for evaluation, then to the MAB Secretariat. In the regional networks, some people are closer to local implementation of biosphere reserves, understand the cultures of the areas and speak local languages. Thus, there would be more regional involvement in the global network (1:21; 14:22). However, some experts also indicated that this might be difficult for political reasons, as some countries might not allow others to evaluate and read their periodic review before they are sent to UNESCO (2:15; 18:16). In addition, the process would require financial and human resources (1:21; 21:13).

The results from the expert interviews provide a good snapshot of the situation at the time of the interviews and show that the selection of experts reflected the reality of the MAB Programme. The ad hoc working group of the MAB ICC, which was established to support the Process of Excellence and Enhancement, arrived at similar conclusions 2019 and recommended that MAB National Committees be established and that biosphere reserves conduct voluntary and informal midterm reviews every five years. This should enable performance monitoring between periodic reviews by national-level authorities. The results also highlighted the importance of regional networks in supporting periodic reviews and facilitating peer-to-peer exchange and support (UNESCO, 2019, pp. 19-21). Future research is urgently needed to investigate the voluntary implementation of five-year reports and the involvement of regional MAB networks.

7.5 Interlinkages with other UNESCO labels

UNESCO manages three different approaches and designations for the conservation and protection of nature: natural World Heritage Sites, Biosphere Reserves and Global Geoparks. These designations may be assigned to individual areas, but they sometimes also overlap. JOB et al. (2017) counted 79 natural World Heritage Sites that were also designated as Biosphere Reserves (cf. JOB et al. 2017: 1712). Moreover, SCHAAF/CLAMOTE RODRIGUES (2016) stated that 109 Biosphere Reserves overlapped with 100 World Heritage Sites, 16 Biosphere Reserves overlapped with 14 Global Geoparks and 15 Global Geoparks overlapped with 13 World Heritage Sites. Two sites even had all three UNESCO designations (cf. SCHAAF/CLAMOTE RODRIGUES 2016: 4f.). Although Biosphere Reserves and World Heritage Sites have different objectives, they can be complementary, as a natural World Heritage Site may comprise the core zone of a biosphere reserve. The buffer and transition zones can enable the sustainable management of the wider surrounding area (cf. Job et al. 2017: 1712f.). Multiple designations also emphasize the international importance of these sites. However, their management also become more complex due to several layers of governance and institutional requirements (cf. MATAR/ANTHONY 2017: 3). PRICE et al. (2010) indicted that the multi-designations were also reflected in additional reporting processes, which still lack a comparative review to increase harmonization and improve information sharing (cf. PRICE et al. 2010: 550). Therefore, the fourth research question addressed the interactions between UNESCO labels: 'What are the interlinkages between the three UNESCO designations: World Heritage Sites, Biosphere Reserves and Global Geoparks? Is it a complementary system of labels, or do they compete against each other?'

The interviewed experts did not have a uniform opinion on whether the labels were complementary or competed with each other. Some recognised that all designations had their own merits and could be complementary, whilst others saw competition for political attention and funding between the labels. Whilst the combination of Biosphere Reserves and World Heritage Sites was favourable, most interviewees believed that Global Geoparks and Biosphere Reserves competed with each other.

A former secretary of the MAB Programme at UNESCO made it clear that the three UNESCO labels for nature conservation were completely different concepts. Biosphere reserves are model sites for sustainable development and use an integrative approach; thus, they are not outstanding and unique areas themselves, unlike World Heritage Sites. Moreover, the new Global Geoparks are distinct in that they do not require a legally protected core zone, as with biosphere reserves, or worldwide outstanding status, as with World Heritage Sites. Thus, the concepts be used in a complementary but sensible way (19:12.2; 19:21). The combination of World Heritage Sites and Global Geoparks can be useful for a unique geological structure. Since Global Geoparks focus on communication and tourism, the combination could provide added value to attract more visitors to the site (27:41; 8:16). Biosphere reserves and World Heritage Sites can also be complementary. The World Heritage Site designation can provide support for the protection of core zones within biosphere reserves. In addition, the protection of World Heritage Sites in biosphere reserves can be integration into the wider landscape through the buffer and transition zones (6:18; 6:29; 21:20; 27:40). One expert even argued that this combination was the best way forward to ensure the long-term protection of the heritage values of the place (14:28). By contrast, many experts saw a competition between biosphere reserves and Global Geoparks; they are incompatible because they have a similar objective: the economic development of the region (27:41; 18:29; 1:28; 14:27; 12:20; 26:19).

In addition to incompatibility and competition between labels in the same areas, many interviewees saw competition over political attention (25:14; 18:30) and funding. All programmes want to show that they have more and better sites. Different labels both within and outside UNESCO compete over funding and do not always work together (13:9; 10:30). Some experts also stated that having only one UNESCO designation would be easier to raise global and local awareness in communities that live around the sites (25:29; 27:40; 29:5; 16:17). Most tourists and locals cannot distinguish between the different UNESCO labels. However, the general public may not need to know this difference as long as they recognize that a site is a UNESCO-designated area (32:23). This is why the use of 'UNESCO Sites' and a single logo for all three labels would be interesting for marketing and raising awareness (25:30; 18:9), also against all other protected area labels, such as national parks, and international designations from other organizations and governments, such as Ramsar sites or Key Biodiversity Areas. Therefore, the main challenge for UNESCO is to develop a clear strategy and management plan for all three nature conservation labels. In addition, it is also crucial for the three UNESCO labels to standardize reporting processes, especially if there will be even more multi-designations for the same site in the future.

Further research is urgently needed on the issue of competition between UNESCO designations, as it has received little attention so far. With the 30% by 2030 target, which will most likely be adopted by the global community at the CBD COP in Kunming in May 2022, UNESCO labels must position themselves better, as they are also in competition with other IUCN nature conservation categories, Ramsar sites and Key Biodiversity Areas. Competition for political recognition and funding between UNESCO labels is detrimental to all programmes and should be avoided at all costs. Better cooperation would also increase acceptance at the local level amongst politicians and communities.

8 Conclusion

Recalling António Guterres's strong statement from the beginning of this thesis at the UN Biodiversity Summit about a war of humanity against nature and his call to live in harmony with nature to avert the worst impacts of climate change and revive biodiversity for the benefit of people and the planet, this thesis provides evidence that, despite many challenges, UNESCO Biosphere Reserves can provide a key framework for greater community participation and regional landscape planning initiatives, therefore serving as a flagship programme for the SDGs and the post-2020 biodiversity agenda.

However, due to a paradigm shift in the biosphere reserve concept after the Seville conference in 1995, there has been an increasing gap in quality between pre- and post-Seville sites, which justified the adoption of quality enhancement strategies: the Exit Strategy in 2013 and, later, the Process of Excellence and Enhancement. The empirical data generated by the 31 expert interviews conducted in 21 countries for this study and crucial UNESCO MAB governance bodies provided new insights on the diversity of the World Network of Biosphere Reserves and different approaches to quality management and the national realization and elaboration of periodic reviews. The four case study countries with the most voluntary withdrawals of biosphere reserves – the United States of America, Australia, Austria and Bulgaria –showed that the Exit Strategy contributed to the streamlining and quality enhancement of the World Network after the paradigm change. All these nations designated biosphere reserves early on in the MAB Programme; thus, voluntarily withdrawn sites after the Exit Strategy were mainly strict nature conservation areas without any human settlements. Only post-Seville sites remain.

Although the inconsistency of many first-generation biosphere reserves with the Statutory Framework was known, only the political pressure occasioned by the implementation of the Exit Strategy led many member states to take action to withdraw sites. However, the Exit Strategy's effects on member states are not limited to countries with voluntary withdrawals. Other member states considered in this thesis, such as China, Ghana, Indonesia, Japan, Kenya, the Republic of Korea and Slovakia, had major problems with first-generation biosphere reserves and needed to make changes to ensure that they conformed to the post-Seville criteria. One finding was that challenges in the implementation of the Statutory Framework, rezonation and the preparation of periodic reviews not only applied to individual countries; all member states with first-generation biosphere reserves had similar problems. Even within states, the situation was very heterogeneous, and there were significant differences in the implementation of biosphere reserves.

All interviewed experts agreed that the quality enhancement strategies were timely and fundamental to improving the credibility and coherence of the MAB Programme. Although there is still an imbalance in the quality of implementation between different biosphere reserves after the Exit Strategy, old, non-functioning biosphere reserves were adapted, or member states were encouraged to voluntarily withdraw them from the network. Many UNESCO member states have become more involved in the MAB Programme, which has led to a rejuvenation of the biosphere reserve network in many countries.

The greatest criticism of the Exit Strategy was its name. Many interviewees felt that the word ‘exit’ has negative connotations and should not be a goal of a strategy in the World Network. Therefore, it was crucial to rename the strategy, which eventually became the Process of Excellence and Enhancement.

The results on the global impact of quality enhancement strategies for UNESCO Biosphere Reserves and their interlinkages with other UNESCO labels represent only a snapshot from 2018. As there are no methodologically comparable analyses from an earlier period, this study is the first global analysis of the Exit Strategy’s effects on the World Network of Biosphere Reserves and national implementation. The geopolitical distribution of interviewees was achieved, and interviews were conducted in each UN region through mainly in-person interactions. To achieve true representativeness, more research is needed in member states in the Global South in the future. Even if these countries have not been the most active and vocal representatives of the MAB ICC to date, they are important in terms of total number of member states, and regional groups in South America and Africa have increased in strength since the implementation of the Exit Strategy. It is of particular importance for the MAB Programme to commission a larger-scale study after the completion of the Exit Strategy in 2021 or 2022 to present impacts and recommendations in member states other than the 21 case study countries considered in this thesis. Given that the MAB Programme is a scientific programme within UNESCO, evaluation studies should focus not only on individual biosphere reserves but MAB National Committees and international MAB governance bodies. In the process, lessons can be learned from the implementation of the Exit Strategy, and a stronger MAB Programme could take the lead in implementing the 2030 Agenda.

The interview experience was much better in onsite and face-to-face conversations than via telephone; telephone interviews were shorter, less intensive and suffered from connection problems. A higher level of trust could be established in onsite interviews, which made it easier to address even the most sensitive and delicate issues. After several interviews, additional information was also provided after the recording device was turned off and during more relaxed conversations over coffee or lunch. The seriousness of the study and the importance of the interviews was underpinned by the researcher’s travel to the case study countries, which was appreciated by interviewees and led to greater participation. During the face-to-face interviews, significant literature and documents were also provided.

Whilst overall feedback on the quality enhancement strategies was positive, the global assessment also revealed many outstanding issues that should be addressed by UNESCO and the MAB Programme:

- Renaming biosphere reserves to ‘biosphere landscapes’ or ‘Man and the Biosphere Sites’ or harmonizing existing variations in terminology should be seriously discussed by the MAB ICC in a timely manner to foster a unified brand and increase the visibility of the World Network of Biosphere Reserves. Whilst MAB governing bodies allow member states to use their own terminology for biosphere reserves, the results of this study show that only half of the 21 case study countries considered in this thesis used the term ‘biosphere reserve’. The word 'reserve' is problematic for many member states; therefore, variations included ‘biosphere area’, ‘biosphere park’, ‘biosphere’ and ‘Eco Park’. UNESCO Biosphere Reserves’ competition over global awareness and recognition with other protected area categories and international designations should not be exacerbated by the fact that terminology varies around the world.

- Strengthening MAB National Committees is key to successful quality management within the network and thus the credibility and coherence of the entire programme. It is important to note that member states are responsible for fulfilling obligations towards the UNESCO MAB Programme in terms of quality management and the completion of periodic reviews. In most case study countries, periodic reviews were conducted in close cooperation with biosphere reserve managers and the MAB National Committee. Thus, the MAB Secretariat must prioritize the creation and strengthening of MAB National Committees; for example, two case study countries in Latin America still do not have one. Positive examples include China and Germany, where MAB National Committees visit biosphere reserves prior to the 10-year evaluation period and provide additional recommendations to the MAB Secretariat. National implementation and enforcement significantly influence the quality of biosphere reserves and are therefore essential for the credibility and coherence of the World Network.
- The focus for future actions in biosphere reserves should be primarily on the buffer and transition zones. Since biosphere reserves must always be created around legally protected core zones, knowledge about the management of core zones is already available. Conflict mediation over the use of natural resources and lands and decision making between the main stakeholders in the buffer and transition zones is the key task of new management staff, also to counterbalance classical nature conservation within the core zone. This could also increase local acceptance of biosphere reserves and facilitate their role as model regions for sustainable development.
- The reporting period for the periodic review process should be reduced and harmonised. Overall, 73% of the experts surveyed in this thesis were in favour of shortening the reporting cycle. Thus, the current 10-year reporting cycle should be adjusted. This decision should take into account other reporting cycles, at least for the two other UNESCO designations, to create synergies and avoid creating more work for sites with multiple designations. A simpler version of the questionnaire with rubrics that people could assign scores to might enable more regular updates. With the completion of the Exit Strategy, The MAB IAC should have less work reviewing periodic reviews that have been pending for decades and could therefore cope with a more frequent reporting cycle.
- Developing a clear strategy and joint management plan for all three nature conservation designations within UNESCO. Competition for political recognition and funding between the UNESCO labels is detrimental to all programmes and should be avoided at all costs. Joint activities and better cooperation would increase acceptance at the local level amongst politicians and local communities. Some experts even stated that having only one UNESCO designation would facilitate global and local awareness amongst communities that live around the sites. Thus, UNESCO should discuss the use of a single label, such as 'UNESCO Sites', for all three designations to enable common marketing and raise awareness.

- New funding sources should be secured for the MAB Secretariat and MAB Programme to manage the increasing number of new biosphere reserves, especially given the key role that the MAB Programme can play in the 2030 Agenda and achieving the SDGs. This funding should be used to keep track of the growing number of periodic reviews, support quality management procedures within the MAB IAC and MAB National Committees and facilitate national participatory nomination processes for new sites. Given increasing global awareness of nature-based solutions for future development, intergovernmental conventions and agreements for combating climate change and biodiversity loss and a pledge of USD 5 billion from nine philanthropic foundations to fund the protection of 30% of land and marine areas by the end of the decade, new funding opportunities should be leveraged by the MAB Programme to strengthen global recognition of sites in the World Network of Biosphere Reserves as model regions for sustainable development.

Bibliography

ABDEL HAMEED, S. (1998). Biosphere Reserves in the Sudan. *Wildlife and Nature, 14*(2), 18-31. https://www.fao.org/3/ak045b/ak045b.pdf

ADAMS, W. M. (2004). *Against Extinction: The Story of Conservation.* Routledge. https://doi.org/10.4324/9781849770415

ADAMS, W. M., & HUTTON, J. (2007). People, Parks and Poverty: Political Ecology and Biodiversity Conservation. *Conservation and Society, 5*(2), 147-183. https://www.jstor.org/stable/26392879

AMEND, S., & AMEND, T. (1995): *National Parks without People? The South American Experience.* IUCN. https://portals.iucn.org/library/node/6891

ANDAM, K. S., FERRARO, P. J., PFAFF, A., SANCHEZ-AZOFEIFA, G. A., & ROBALINO, J. A. (2008). Measuring the effectiveness of protected area networks in reducing deforestation. *Proceedings of the National Academy of Sciences of the United States of America (PNAS), 105*(42), 16089-16094. https://doi.org/10.1073/pnas.0800437105

ANDERSON, D., & GROVE, R. H. (1987). *Conservation in Africa: People, policies and practice.* Cambridge University Press.

ANTHAMATTEN, P., & HAZEN, H. (2015). Changes in the Global Distribution of Protected Areas, 2003-2012. *The Professional Geographer, 67*(2), 195-203. https://doi.org/10.1080/00330124.2014.921014

ASHONG, S. N .A., ASOMANING, W. A., MENSAH, A., TETTEH, E., & AGYEMAN, D. Y. (2013). Natural Resources, People and Livelihoods in the Songor Biosphere Reserve. In: R. Pool-Stanvliet & M. Clüsener-Godt (Eds.), *AfriMAB. Biosphere Reserves in Sub-Saharan Africa: Showcasing Sustainable Development.* UNESCO, 123-140. https://unesdoc.unesco.org/ark:/48223/pf0000227316

BATISSE, M. (1982). The Biosphere Reserve: A Tool For Environmental Conservation and Management. *Environmental Conservation, 9*(2), 101-111. https://www.jstor.org/stable/44517328

BATISSE, M. (1986). Developing and focusing the biosphere reserve concept. *Nature and Resources, 22,* 1-10.

BOGNER, A., LITTING, B., & MENZ, W. (2009). Introduction: Expert Interviews - An Introduction to a New Methodological Debate. In: A. Bogner, B. Litting, & W. Menz (Eds.), *Interviewing Experts.* Palgrave Macmillan, 1-13. https://doi.org/10.1057/9780230244276_1

BOGNER, A., LITTING, B., & MENZ, W (2014). *Interviews mit Experten. Eine praxisorientierte Einführung.* Springer VS.

BORGES, M. A., CARBONE, G., BUSHELL, R., & JAEGER, T. (2011). *Sustainable tourism and natural World Heritage: Priorities for action.* IUCN. https://portals.iucn.org/library/node/9789

BORRINI-FEYERABEND, G., DUDLEY, N., JAEGER, T., LASSEN, B., PATHAK BROOME, N., PHILLIPS, A., & SANDWITH, T. (2013). *Governance of Protected Areas. From understanding to action.* Best Practice Protected Area Guidelines Series No. 20. IUCN. https://portals.iucn.org/library/sites/library/files/documents/PAG-020.pdf

BOTHA, N., JOB, H., KIMARIO, F. (2021). Potential and challenges of the Serengeti-Ngorongoro Biosphere Reserve, Tanzania. *Eco.mont, 13,* 27-37. https://doi.org/10.1553/eco.mont-13-sis27

BRENNER, L., JOB, H. (2022). Reviewing the participatory management of UNESCO Biosphere Reserves: What do we miss by ignoring local academic knowledge in Mexico? *Ambio,* 51, 1726-1738. https://doi.org/10.1007/s13280-021-01672-1

BRIDGEWATER, P., PHILIPPS, A., GREEN, M., & AMOS, B. (1996). *Biosphere Reserves and the IUCN System of Protected Area Management Categories.* Australian Nature Conservation Agency, IUCN, UNESCO (MAB). https://portals.iucn.org/library/node/7592

BRIDGEWATER, P. (2016). The Man and Biosphere programme of UNESCO: rambunctious child of the sixties, but was the promise fulfilled? *Current Opinion in Environmental Sustainability, 19,* 1-6. https://doi.org/10.1016/j.cosust.2015.08.009

BRIDGEWATER, P. (2020). The four stages of biosphere reserves in Australia: A retro- and prospective. In: M. Reed & M. Price (Ed.): *UNESCO Biosphere Reserves: Supporting Biocultural Diversity, Sustainability and Society.* Routledge, 228-238.

BROCKINGTON, D., & SCHMIDT-SOLTAU, K. (2004). The social and environmental impacts of wilderness and development. *Oryx 38,* 140-142. https://doi.org/10.1017/S0030605304000250

BRUNCKHORST, D. (2001). Building capital through bioregional planning and biosphere reserves. *Ethics in Science and Environmental Politics ESEP,* 19-32.

BROCKINGTON, D., & IGOE, J. (2006). Eviction for Conservation: A Global Overview. *Conservation and Society, 4*(3), 424-470. https://www.jstor.org/stable/26396619

BUCKLEY, R. (2004). The Effects of World Heritage Listings on Tourism to Australian National Parks. *Journal of Sustainable Tourism, 12*(1), 70-84. https://doi.org/10.1080/09669580408667225

CARRUTHERS, J. (1995): *The Kruger National Park: A Social and Political History.* University of Natal Press.

CASTREE, N. (2014): *Making sense of nature. Representation, politics and democracy.* Routledge.

CASTREE, N. (2017) Nature. In: D. Richardson, N. Castree, M. Goodchild, A. Kobayashi, W. Liu, R. Marston (Eds.): *The International Encyclopedia of Geography*. Wiley. 1-26. https://doi.org/10.1002/9781118786352

CHAPE, S., HARRISON, J., SPALDING, M., & LYSENKO, I. (2005). Measuring the extent and effectiveness of protected areas as an indicator for meeting global biodiversity targets. *Philosophical Transactions of the Royal Society B, 360*(1454), 443-455. https://doi.org/10.1098/rstb.2004.1592

CHRISTMANN, G. B. (2009). Expert Interviews on the Telephone: A Difficult Undertaking. In: Bogner, A., Litting, B., Menz, W (Ed.), *Interviewing Experts*. Palgrave Macmillan, 157-183. https://doi.org/10.1057/9780230244276_8

CLÜSENER-GODT, M. (2012). Spaniens innovative Nutzung des Konzepts der Biosphärenreservate national und international. *Natur und Landschaft, 87*(9/10), 435-440. https://www.natur-und-landschaft.de/de/magazin?id=48

COAD, L., WATSON, J. E.M., GELDMANN, J., BURGESS, N. D., LEVERINGTON, F., HOCKINGS, M., KNIGHTS, K., & DI MARCO, M. (2019). Widespread shortfalls in protected area resourcing undermine efforts to conserve biodiversity. *Frontiers in Ecology and the Environment, 17*(5), 259-264. https://doi.org/10.1002/fee.2042

COETZER, K., WITKOWSKI, E., & ERASMUS, B. (2014). Reviewing Biosphere Reserves globally: effective conservation action or bureaucratic label? *Biological Reviews, 89*(1), 82-104. https://doi.org/10.1111/brv.12044

CONRADIN, K., ENGESSER, M., & WIESMANN, U. (2015). Four decades of World Natural Heritage - how changing protected area values influence the UNESCO label. *DIE ERDE, 146*(1), 34-46. https://doi.org/10.12854/erde-146-4

CONRADIN, K., CHIYUMBA, M., KITEME, B., MWAURA, S. N., LIECHTI, K. (2014). World Natural Heritage sites and regional development - the cases of Mt Kenya and Mt Kilimanjaro. *eco.mont, 6*(1), 45-48. http://dx.doi.org/10.1553/ecomont-6-1s45

CONVENTION ON BIOLOGICAL DIVERSITY (CBD). (2010). *Strategic Plan for Biodiversity 2011-2020 and the Aichi Biodiversity Targets*. CBD.

CONVENTION ON BIOLOGICAL DIVERSITY (CBD). (2020). *Update of the Zero Draft of the post-2020 Global Biodiversity Framework*. CBD.

DEGUIGNET, M., ARNELL, A., JUFFE-BIGNOLI, D., SHI, Y., BINGHAM, H., MACSHARRY, B., & KINGSTON, N. (2017). Measuring the extent of overlaps in protected area designations. *PLOS one, 12*(11), 1-17. https://doi.org/10.1371/journal.pone.0188681

DINERSTEIN, E., VYNNE C., SALA, E., JOSHI, A. R., FERNANDO, S., LOVEJOY, T. E., MAYORGA, J., OLSON, D., ASNER, G., BAILLIE, J. E. M., BURGEES, N. D., BURKART, K., NOSS, R. F.,, ZHANG Y. P., BACCINI, A., BIRCH, T., HAHN, N., JOPPA, L. N., & WIKRAMANAYAKE, E.

(2019). A Global Deal for Nature: Guiding principles, milestones, and targets. *Science Advances 5*, 1-17. https://doi.org/10.1126/sciadv.aaw2869

Ding, W., & Qunli, H. (2020). The Man and the Biosphere Programme in China: Sustaining the community of live over four decades. In: M. Reed & M. Price (Eds.), *UNESCO Biosphere Reserves: Supporting Biocultural Diversity, Sustainability and Society.* Routledge, 213-227. https://doi.org/10.4324/9780429428746

Dudley, N. (2008): *Guidelines for Applying Protected Area Management Categories.* IUCN. https://portals.iucn.org/library/sites/library/files/documents/PAG-021.pdf

Dunn, K. M. (2006). Interviewing. In: I. Hay (Ed.): *Qualitative Research Methods in Human Geography.* Oxford University Press.

Dyer, M. I., & Holland, M. M. (1988). Unesco's Man and the Biosphere Program. *Bioscience, 38*(9), 635-641. https://www.jstor.org/stable/1310830

Dyer, M. I., & Holland, M. M. (1991). The Biosphere-Reserve Concept: Needs for a Network Design. *BioScience, 41*(5), 319-325. https://www.jstor.org/stable/1311585

Eder, W. (1999). UNESCO GEOPARKS - A new initiative for protection and sustainable development of the Earth's heritage. *Neues Jahrbuch für Geologie und Paläontologie - Abhandlungen, 214*(1), 353-358. https://doi.org/10.1127/njgpa/214/1999/353

Elbakidze, M., Hahn, T., Mauerhofer, V., Angelstam, P., & Axelsson, R. (2013). Legal Framework for Biosphere Reserves as Learning Sites for Sustainable Development: A Comparative Analysis of Ukraine and Sweden. *AMBIO, 42,* 174-187. https://doi.org/10.1007/s13280-012-0373-3

Emerson, R. W. (1836). *Nature.* James Munroe and Company.

Engels, B., Manz, K., & Job-Hoben, B. (2011). Natural World Heritage Sites and tourism: Challenges and opportunities. *Natur und Landschaft 86*(12), 527-533.

Franklin, J. (1977). The Biosphere Reserve Program in the United States: A program has been developed to select key sites for environmental research and monitoring. *Science, 195*(4275), 262-267. https://doi.org/10.1126/science.195.4275.262

Geldmann, J., Coad, L., Barnes, M., Craigie, I. D., Hockings, M., Knights, K., Leverington, F., Cuadros, I. C., Zamora, C., Woodley, S., & Burgess, N. D. (2015). Changes in protected area management effectiveness over time: A global analysis. *Biological Conservation, 191*, 692-699. https://doi.org/10.1016/j.biocon.2015.08.029

Geldmann, J., Manica, A., Burgess, N. D., Coad, L., & Balmford, A. (2019). A global-level assessment of the effectiveness of protected areas at resisting anthropogenic pressures. *PNAS, 116*(46), 23209-23215. https://doi.org/10.1073/pnas.1908221116

German Federal Agency for Nature Conservation (BfN). (2020). *The German MAB National Committee.* Retrieved December 14, 2020, from https://www.bfn.de/en/

activities/international-nature-conservation/international-agreements-and-programmes/steckbriefe-internationales/man-and-the-biosphere/mab-committee.html

GHANA MAN AND THE BIOSPHERE (MAB) NATIONAL COMMITTEE. (2018). *Ghana Action Plan (2018-2015) for Implementation of UNESCO Man and the Biosphere (MAB) Programme*. Ghana MAB National Committee.

GHANA MAN AND THE BIOSPHERE (MAB) NATIONAL COMMITTEE. (2019). *Report on progress of implementation of the Lima Action Plan (2016 - 2025) presented to the 31st session of the MAB Council, Paris-France, June 2019*. UNESCO. https://en.unesco.org/sites/default/files/ghana_national_report.pdf

GHANA NATIONAL COMMISSION FOR UNESCO. (2019): *2019 Annual Report*. Ghana National Commission for UNESCO.

GILBERT, V. (2016). *Biosphere Reserves and Sovereignty: A Story for the United States Man and Biosphere (MAB) Community*.

GLÄSER, J., & LAUDEL, G. (2010). *Experteninterviews und qualitative Inhaltsanalyse: als Instrumente rekonstruierender Untersuchungen* (4th ed.). VS, Verlag für Sozialwissenschaften.

GÓMEZ SAL, A. G. (2018). La Red Española de Reserveas de la Biosfera. Una historia de éxito ante el reto de incorporar nuevas funciones. *Ambienta: La revista del Ministerio de Medio Ambiente, 124*, 68-81.

GRAY, C. L., HILL, S. L. L., NEWBOLD, T., HUDSON, L. N., BÖRGER, L., CONTU, S., HOSKINS, A. J., FERRIER, S., PURVIS, A., SCHARLEMANN, J. P. W. (2016). Local biodiversity is higher inside than outside terrestrial protected areas worldwide. *Nature Communications, 7*(12306), 1-7. https://doi.org/10.1038/ncomms12306

GREGG, W. P. Jr. (1983). MAB and its Biosphere Reserves Project: A new dimension in Global Conservation. *The George Wright Forum, 3*(2), 17-31. https://www.jstor.org/stable/43596998

GRZIMEK, B. (1959). *Serengeti darf nicht sterben: 367000 Tiere suchen einen Staat*. Ullstein.

GROVE, R. H. (1996). *Green imperialism: Colonial expansion, tropical island Edens and the origins of environmentalism, 1600 - 1860*. Cambridge University Press.

GUTERRES, A. (2020, September 30). *Secretary-General's remarks to United Nations Biodiversity Summit [Speech as delivered]*. United Nations. https://www.un.org/sg/en/content/sg/statement/2020-09-30/secretary-generals-remarks-united-nations-biodiversity-summit-delivered

HIGH AMBITION COALITION (HAC) FOR NATURE AND PEOPLE. (2021). *Home*. Retrieved December 29, 2021, from https://www.hacfornatureandpeople.org/home

Haddad, N. M., Brudvig, L. A., Clobert, J., Davies, K. F., Gonzalez, A., Holt, R. D., Lovejoy, T. E., Sexton, J. O., Austin, M. P., Collins, C. D., Cook, W. M., Damschen, E. I., Ewers, R. M., Foster, B. L., Jenkins, C. N., King, A. J., Laurance, W. F., Levey, D. J., Margules, C. R., Melbourne, B. A., Nicholls, A. O., Orrock, J. L., Song, D.-X., & Townshend, J. R. (2015). Habitat fragmentation and its lasting impact on Earth's ecosystems. *Science Advances, 1*(2), 1-9. https://doi.org/10.1126/sciadv.1500052

Hall, C. M., & Piggin, R. (2001). Tourism and World Heritage in OECD Countries. *Tourism Recreation Research, 26*(1), 103-105. https://doi.org/10.1080/02508281.2001.11081182

Harper, D. (n.d.). Etymology of nature. Online Etymology Dictionary. Retrieved November 11, 2020, from https://www.etymonline.com/word/nature

Henriques, M. H., & Brilha, J. B. (2017). UNESCO Global Geoparks: a strategy towards global understanding and sustainability. *Episodes, 40*(4), 349-355. https://doi.org/10.18814/epiiugs/2017/v40i4/017036

Hoffmann, S., Beierkuhnlein, C., Field, R., Provenzale, A., & Chiarucci, A. (2018). Uniqueness of Protected Areas for Conservation Strategies in the European Union. *Scientific Reports Nature, 8*(6445), 1-14. https://doi.org/10.1038/s41598-018-24390-3

Hutton, J., Adams, W. M., & Murombedzi, J. C. (2005). Back to the Barriers? Changing Narratives in Biodiversity Conservation. *Forum for Development Studies, 32*(2), 341-370. https://doi.org/10.1080/08039410.2005.9666319

International Union for Conservation of Nature and Natural Resources (IUCN). (1978). *Categories, objectives and criteria for protected areas.* IUCN. https://portals.iucn.org/library/sites/library/files/documents/1978-003.pdf

International Union for Conservation of Nature (IUCN). (2018). *The impact of IUCN resolutions on international conservation efforts: an overview.* IUCN. https://portals.iucn.org/library/sites/library/files/documents/2018-011-En.pdf

International Union for Conservation of Nature (IUCN). (2019). *About.* Retrieved July 28, 2019, from https://www.iucn.org/about/

Ishwaran, N. (2012). Science in intergovernmental environmental relations: 40 years of UNESCO's Man and the Biosphere (MAB) Programme and its future. *Environmental Development, 1*(1), 91-101. https://doi.org/10.1016/j.envdev.2011.11.001

Ishwaran, N., Persic, A., & Tri, N.H. (2008). Concept and practice: the case of UNESCO biosphere reserves. *Environment and Sustainable Development, 7*(2), 118-131.

Jay, M., & Morad, M. (2009). Conservation and Ecology. In: R. Kitchin & N. Thrift (Eds.), *International Encyclopedia of Human Geography Vol. 2.* Elsevier Science, 259-267.

JUNGMEIER, M., BORSDORF, A., BRAUN, V., HÄRING, V., HAMMER, T., & PICHLER-KOBAN, C. (2021). Pärke, Parks and Reservate - biosphere reserves in Austria, Germany and Switzerland on their way towards Biosphere 4.0? *eco.mont, 13*, 15-26. https://doi.org/10.1553/eco.mont-13-sis15

JOB, H., BECKEN, S., & LANE, B. (2017). Protected Areas in a neoliberal world and the role of tourism in supporting conservation and sustainable development: an assessment of strategic planning, zoning, impact monitoring, and tourism management at natural World Heritage Sites. *Journal of Sustainable Tourism, 25*(12), 1697-1718. https://doi.org/10.1080/09669582.2017.1377432

JOB, H., ENGELBAUER, M., & ENGELS, B. (2019). The portfolio of German biosphere reserves in the light of the Sustainable Development Goals. *Spatial Research and Planning, 77*(1), 57-79. https://doi.org/10.2478/rara-2019-0005

JONES, C. (2008). History of Geoparks. *Geological Society Special Publications, 300*, 273-277. https://doi.org/10.1144/SP300.21

KJELLQVIST, T., RODELA, R., & LEHTILÄ, K. (2020). Meeting the challenge of sustainable development: Analysing the knowledge used to establish Swedish biosphere reserves. In: M. Reed & M. Price (Eds.), *UNESCO Biosphere Reserves: Supporting Biocultural Diversity, Sustainability and Society*. Routledge, 102-113.

KÖCK, G., & ARNBERGER, A. (2017). The Austrian Biosphere Reserves in the light of changing MAB strategy. *eco.mont, 9*, 85-92. https://doi.org/10.1553/eco.mont-9-sis85

KÖCK, G., ARNBERGER, A., & MÖLLER, L. (2020). Agenda 2030 und Lima-Aktionsplan - Anpassungen der *Biosphere Reserves* für die Zukunft. In A. Borsdorf, M. Jungmaier, V. Braun, & K. Heinrich (Eds.), *Biosphäre 4.0*, 61-83. https://doi.org/10.1007/978-3-662-60707-7_3

KOHSAKA, R., & MATSUOKA, H. (2015). Analysis of Japanese Municipalities With Geopark, MAB, and GIAHS Certifications: Quantitative Approach to Official Records With Text-Mining Methods. *SAGE Open*, 1-10. https://doi.org/10.1177%2F2158244015617517

LEUNG, Y.-F., SPENCELEY, A., HVENEGAARD, G., & BUCKLEY, R. (2018): *Tourism and visitor management in protected areas: Guidelines for sustainability*. IUCN. https://portals.iucn.org/library/sites/library/files/documents/PAG-027-En.pdf

LE SAOUT, S., HOFFMANN, M., SHI, Y., HUGHES, A., BERNARD, C., BROOKS, T. M., BERTZKY, B., BUTCHART, S. H. M., STUART, S. N., BADMAN, T., & RODRIGUES, A. S. L. (2013). Protected Areas and Effective Biodiversity Conservation. *Science, 342*, 803-805. https://doi.org/10.1126/science.1239268

LEVERINGTON, F., COSTA, K. L., PAVESE, H., LISLE, A., & HOCKINGS, M. (2010). A Global Analysis of Protected Area Management Effectiveness". In: *Environmental Management, 46*, 685-698. https://doi.org/10.1007/s00267-010-9564-5

MAB National Committee of the Republic of Korea. (2015). *Republic of Korea MAB Strategy and Action Plan 2016-2025.* MAB National Committee of the Republic of Korea.

Mabunda, D., Pienaar, D., & Verhoef, J. (2003). The Kruger National Park: A Century of Management and Research. In: J. T. Du Toit, K. H. Rogers, & H. C. Biggs, (Eds.), *The Kruger Experience: Ecology and Management of Savanna Heterogeneity.* Island Press, 3-21.

Matar, D. A., & Anthony, B. P. (2017). UNESCO Biosphere Reserve management evaluation: where do we stand and what's next. *International Journal of UNESCO Biosphere Reserves, 1*(1), 1-22.

Matar, D. A., & Anthony, B. P. (2018). Implementation of the periodic review requirement in the Arab-MaB Network: lessons for improving biosphere reserve evaluation. *Parks, 24*(1), 7-22. https://www.doi.org/10.2305/IUCN.CH.2018.PARKS-24-1.en

Mathevet, R., & Cibien, C. (2020). The French biosphere reserves: Looking for ecological solidarity and stewardship. In: M. Reed & M. Price (Eds.), *UNESCO Biosphere Reserves: Supporting Biocultural Diversity, Sustainability and Society.* Routledge, 114-124.

Matsuda, H., Yumoto, T., Okano, T., Tetsuka, K., Fujimaki, A., & Shioya, K. (2015). An extension plan of Yakushima Biosphere Reserve as a case study of consensus building of islanders. *Journal of Ecology and Environment, 38*(2), 241-247. https://doi.org/10.5141/ecoenv.2015.024

Matsuda, H., Nakamura, S., & Sato, T. (2020). Transdisciplinary approaches for the reactivation of Japanese biosphere reserves. In: M. Reed & M. Price (Eds.), *UNESCO Biosphere Reserves: Supporting Biocultural Diversity, Sustainability and Society.* Routledge, 190-200.

Matysek, K. A., Stratford, E., & Kriwoken, L. K. (2006). The UNESCO Biosphere Reserve Program in Australia: constraints and opportunities for localized sustainable development. *The Canadian Geographer, 50*(1), 85-100. https://doi.org/10.1111/j.0008-3658.2006.00128.x

Maxwell, S. L., Cazalis, V., Dudley, N., Hoffmann, M., Rodrigues, A. S. L., Stolton, S., Visconti, P., Woodley, S., Kingston, N., Lewis, E., Maron, M., Strassburg, B. B. N., Wenger, A., Jonas, H. D., Venter, O., & Watson, J. E. M. (2020). Area-based conservation in the twenty-first century. *Nature, 586*, 217-227. https://doi.org/10.1038/s41586-020-2773-z

Mayerl, D. (2005). The Network of Biosphere Reserves in Germany In: German MAB National Committee (Eds.), *Full of Life.* Springer, 23-31. https://link.springer.com/chapter/10.1007/3-540-25815-9_4

MAYRING, P. (2002). *Einführung in die qualitative Sozialforschung: Eine Anleitung zu qualitativem Denken*. Beltz.

MERCER, D., & HYMAN, G. (2009). Unfulfilled Promise: the case of the Mornington Peninsula and Western Port Biosphere Reserve, Australia. *Australian Geographer, 40*(4), 409-427. https://doi.org/10.1080/00049180903312638

MEUSER, M., & NAGEL, U. (2009). The Expert Interview and Changes in Knowledge Production. In: A. Bogner, B. Littig, & W. Menz (Eds.), *Interviewing Experts*. Palgrave Macmillan, 17-42. https://doi.org/10.1057/9780230244276_2

MILGROOM, J., & SPIERENBURG, M. (2008). Induced volition: Resettlement from the Limpopo National Park, Mozambique. *Journal of Contemporary African Studies, 26*(4), 435-448. https://doi.org/10.1080/02589000802482021

MILLENNIUM ECOSYSTEM ASSESSMENT. (2005). *Ecosystems and Human Well-being: Synthesis.* Island Press. https://www.unep.org/resources/report/ecosystem-and-human-well-being-synthesis

MILNE, R., & WAUGH, J. (1994). North America. In: J. McNeely, J. Harrison, & P. Dingwall (Eds.), *Protecting Nature: Regional Reviews of Protected Areas*. IUCN, 277-300.

MINISTERIUM FÜR UMWELT, KLIMA UND ENERGIEWIRTSCHAFT (MUKE) BADEN-WÜRTTEMBERG. (2020). *Biosphärengebiete.* https://um.baden-wuerttemberg.de/de/umwelt-natur/naturschutz/schutzgebiete/biosphaerengebiete/

MISHRA, H. (1994). South and Southeast Asia. In: J. McNeely, J. Harrison, & P. Dingwall (Eds.), *Protecting Nature: Regional Reviews of Protected Areas*. IUCN, 177-204.

NAUBER, J. O., & POKORNY, D. (1994). Establishment of biosphere reserves in Germany. *Parks, 4*(3), 27-35.

NAUGHTON-TREVES, L., ALVAREZ-BERRÍOS, N., BRANDON, K., BRUNER, A., HOLLAND, M. B., PONCE, C., SAENZ, M., SUAREZ, L., & TREVES, A. (2006). Expanding protected areas and incorporating human resource use: a study of 15 forest parks in Ecuador and Peru. *Sustainability: Science, Practice and Policy, 2*(2), 32-44. https://doi.org/10.1080/15487733.2006.11907983

NEUMANN, R. P. (1996). Dukes, earls, and ersatz Edens: aristocratic nature preservations in colonial Africa. *Environment and Planning D: Society and Space, 14*(1), 79-98. https://doi.org/10.1068%2Fd140079

NEUMANN, R. P. (1998). *Imposing Wilderness: Struggles over livelihood and nature preservation in Africa*. University of California Press.

NEUMANN, R. P. (2009). Biodiversity. In: R. Kitchin & N. Thrift (Eds.), *International Encyclopedia of Human Geography Vol. 1*. Elsevier Science, 308-313.

NOLTE, B. (2016). Can Tourism Promote Regional Development in Protected Areas? Case Studies from the Biosphere Reserves Slovensky Kras and Polana, Slovakia. In: I. Mose (Ed.), *Protected Areas and Regional Development in Europe: Towards a New Model for the 21st Century.* Ashgate, pp. 55-74.

NORMAN, E. S., & CARR, D. (2009). Rio Summit. In: R. Kitchin & N. Thrift (Eds.), *International Encyclopedia of Human Geography Vol. 2.* Elsevier Science, 406-411.

OLDEKOP, J. A., HOLMES, G., HARRIS, W. E., & EVANS, K. L. (2016). A global assessment of the social and conservation outcomes of protected areas. *Conservation Biology, 30*(1), 133-141. https://doi.org/10.1111/cobi.12568

OLINDO, P., & MBAELELE, M. (1994): Sub-Saharan Africa. In: J. McNeely, J. Harrison, & P. Dingwall (Eds.), *Protecting Nature: Regional Reviews of Protected Areas.* IUCN, 43-72.

OSTERGREN, D. (2001). An Organic Act after a Century of Protection: The Context, Content, and Implications of the 1995 Russian Federation Law on Specifically Protected Natural Areas. *Natural Resources Journal, 41*(1), 125-152. https://www.jstor.org/stable/24888550

ÖSTERREICHISCHES NATIONALKOMITEE MAB PROGRAMME. (2021). *50 Jahre MAB-Programm („Man and the Biosphere"): Eine Erfolgsgeschichte der UNESCO.* Retrieved July 21, 2021, from http://www.biosphaerenparks.at/index.php/de/blog/105-50-jahre-mab-programm-%E2%80%9Eman-and-the-biosphere%E2%80%9C-eine-erfolgsgeschichte-der-unesco

ÖSTERREICHISCHE UNESCO-KOMMISSION. (2020). *UNESCO Biosphärenparks.* Retrieved December 19, 2020, from https://www.unesco.at/wissenschaft/der-mensch-und-die-biosphaere/unesco-biosphaerenparks

PALOMO, I., MONTES, C., MARTÍN-LÓPEZ, B., GONZÁLEZ, J. A., GARCÍA-LLORENTE, M., ALCORLO, P., & GARCÍA MORA, M. R. (2014). Incorporating the Social-Ecological Approach in Protected Areas in the Anthropocene. *BioSience, 63*(3), 181-191. https://doi.org/10.1093/biosci/bit033

PATZAK, M., & EDER, W. (1998). 'UNESCO GEOPARK': A new Programme - A new UNESCO label. *Geologica Balcanica, 28*(3-4), 33-35.

PFADENHAUER, M. (2009). At Eye Level: The Expert Interview - a Talk between Expert and Quasi-expert. In: A. Bogner, B. Litting, & W. Menz (Eds.), *Interviewing Experts.* Palgrave Macmillan, 81-97. https://doi.org/10.1057/9780230244276_4

PHILIPPS, W. (2004). The history of the international system of protected area management categories. *Parks, 14*(3), 4-14.

POFF, B. (1995). Introducing a new website for the USA Man and Biosphere Program (MAB). *Hydrology and Water Resources in Arizona and the Southwest, 35*, 63-68. http://hdl.handle.net/10150/296615

POIRIER, R., & OSTERGREN, D. (2002). Evicting People from Nature: Indigenous Land Rights and National Parks in Australia, Russia, and the United States. *Natural Resources Journal, 42*(2), 331-351.

PRICE, M. F. (2002). The periodic review of biosphere reserves: a mechanism to foster sites of excellence for conservation and sustainable development. *Environmental Science & Policy 5*(1), 13-18. https://doi.org/10.1016/S1462-9011(02)00021-7

PRICE, M. F. (2017). The re-territorialisation of Biosphere Reserves: The case of Wester Ross, Northwest Scotland. *Environmental Science & Policy, 72*, 30-40. https://doi.org/10.1016/j.envsci.2017.02.002

PRICE, M. F. (2020). The evolution of the biosphere reserve network in the United Kingdom. In: M. Reed & M. Price (Eds.), *UNESCO Biosphere Reserves: Supporting Biocultural Diversity, Sustainability and Society*. Routledge, 89-101.

PRICE, M. F., PARK, J. J., & BOUAMRANE, M. (2010). Reporting progress on internationally designated sites: The periodic review of biosphere reserves. *Environmental Science & Policy, 13*, 549-557. https://doi.org/10.1016/j.envsci.2010.06.005

PRYDE, P. R. (1997). Post-Soviet Development and Status of Russian Nature Reserves. *Post-Soviet Geography and Economics, 38*(2), 63-80. https://doi.org/10.1080/10889388.1997.10641037

PURWANTO, Y., NUGROHO, H., SETIAWAN ACHMADI, A. S., & MUNAWAROH, E. (2020). Biosphere Reserve Concept Implementation for Creating Sustainability. *Journal of Tropical Ethnobiology, 3*(1), 43-56. https://doi.org/10.46359/jte.v3i1.7

RED ESPAÑOLA DE RESERVAS DE LA BIOSFERA (RERB). (2020). *Mapa de Reservas: Las Reservas de la Biosfera Españolas*. Retrieved December 19, 2020, from http://rerb.oapn.es/red-espanola-de-reservas-de-la-biosfera/reservas-de-la-biosfera-espanolas/mapa

REED, M. G. (2016). Conservation (In)Action: Renewing the Relevance of UNESCO Biosphere Reserves. *Conservation Letters, 9*(6), 448-456. https://doi.org/10.1111/conl.12275

ROBERTSON, P. (2001). Sudan. In: L. D. C. Fishpool & M. I. Evans (Eds.), *Important Bird Areas in Africa and associated islands*. Pisces Publications & BirdLife International, 877-890.

RODRIGUES, A. S. L., ANDELMAN, S. J., BAKARR, M. I., BOITANI, L., BROOKS, T. M., COWLING, R. M., FISHPOOL, L. D. C., DA FONSECA, G. A. B., GASTON, K. J., HOFFMANN, M., LONG, J. S., MARQUET, P. A., PILGRIM, J. D., PRESSEY, R. L., SCHIPPER, J., SECHREST, W., STUART, S. N., UNDERHILL, L. G., WALLER, R. W., WATTS, M. E. J., YAN, X. (2004).

Effectiveness of the global protected area network in representing species diversity. *Nature, 428*, 640-643. https://doi.org/10.1038/nature02422

SALU, E. (2013). Protecting Bia Biosphere Reserve for Improved Biodiversity Conservation in Ghana. In: R. Pool-Stanvliet & M. Clüsener-Godt (Eds.), *AfriMAB. Biosphere Reserves in Sub-Saharan Africa: Showcasing Sustainable Development.* UNESCO, 73-89.

SANTOS-MARTÍNEZ, A. S., HINOJOSA, S., & SIERRA ROZO, O. (2013). Proceso y avance hacia la Sostenibilidad Ambiental: La Reserva de la Biosfera Seaflower, en el Caribe Colombiano. *Cuadernos del Caribe, 13*, 1-23.

SAUNDERS, D. A., HOBBS, R. J., & MARGULES, C. R. (1991). Biological Consequences of Ecosystem Fragmentation: A Review. *Conservation Biology, 5*(1), 18-32. https://doi.org/10.1111/j.1523-1739.1991.tb00384.x

SCHAAF, T., & CLAMOTE RODRIGUES, D. (2016). *Managing MIDAs: Harmonising the management of Multi-Internationally Designated Areas: Ramsar Sites, World Heritage sites, Biosphere Reserves and UNESCO Global Geoparks.* IUCN. https://portals.iucn.org/library/sites/library/files/documents/2016-033.pdf

SHEPARD JR., G. H., RUMMENHOELLER, K., OHL-SCHACHERER, J., & YU, D. W. (2010). Trouble in Paradise: Indigenous Populations, Anthropological Policies, and Biodiversity Conservation in Manu National Park, Peru. *Journal of Sustainable Forestry, 29*(2-4), 252-301. https://doi.org/10.1080/10549810903548153

SOBISEVICH, A. V., & SNYTKO, V. A. (2020). To the history of the creation of biosphere reserve and natural parks in the Soviet Union. *IOP Conference Series: Earth and Environmental Science, 579*, 1-5. https://doi.org/10.1088/1755-1315/579/1/012076

SOKOLOV, V. (1981). The Biosphere Reserve Concept in the USSR. *Ambio, 10*(2/3), 71-101. https://www.jstor.org/stable/4312649

STOLL-KLEEMANN, S., DE LA VEGA-LEINERT, A. C., & SCHULTZ, L. (2010). The role of community participation in the effectiveness of UNESCO Biosphere Reserve management: evidence and reflections from two parallel global surveys. *Environmental Conservation 37*(3), 227-238. https://doi.org/10.1017/S037689291000038X

STOLL-KLEEMANN, S., & O'RIORDAN, T. (2018) Biosphere Reserves in the Anthropocene. In: D. DELLASALA & M. GOLDSTEIN (Eds.), *The Encyclopedia of the Anthropocene* 3. Elsevier, 347-353.

SU, Y.-W., & LIN, H.-L. (2014). Analysis of international tourist arrivals worldwide: The role of world heritages sites. *Tourism Management 40*, 46-58. http://doi.org/10.1016/j.tourman.2013.04.005

SUDARMONOWATI, E. (2019). Implementation of Response to Lima Action Plan in Indonesia: Strategy and Progress in Relation to SDGs. *IOP Conference Series: Earth and*

Environmental Science, *298*(012020), 1-12. https://doi.org/10.1088/1755-1315/298/1/012020

SUESS, E. (1875): *Die Entstehung der Alpen*. Braumüller.

SULAYEM, M., SALEH, M., DEAN, F., & DRUCKER, G. (1994). North Africa and the Middle East. In: J. McNeely, J. Harrison, & P. Dingwall (Eds.), *Protecting Nature: Regional Reviews of Protected Areas*. IUCN, 73-100.

UDVARDY, M. F. (1975): *A Classification of the Biogeographical Provinces of the World. Prepared as a Contribution to UNESCO's Man and the Biosphere Programme Project No. 8*. IUCN Occasional Paper, No. 18. IUCN.

UNITED NATIONS ENVIRONMENT PROGRAMME WORLD CONSERVATION MONITORING CENTRE (UNEP-WCMC). (2020a). *Protected Area Profile for Africa from the World Database of Protected Areas, November 2020*. Retrieved December 27, 2020, from https://www.protectedplanet.net

UNITED NATIONS ENVIRONMENT PROGRAMME WORLD CONSERVATION MONITORING CENTRE (UNEP-WCMC). (2020b). *Protected Area Profile for Asia & Pacific from the World Database of Protected Areas, November 2020*. Retrieved December 27, 2020, from https://www.protectedplanet.net

UNITED NATIONS ENVIRONMENT PROGRAMME WORLD CONSERVATION MONITORING CENTRE (UNEP-WCMC). (2020c). *Protected Area Profile for Europe from the World Database of Protected Areas, November 2020*. Retrieved December 27, 2020, from https://www.protectedplanet.net

UNITED NATIONS ENVIRONMENT PROGRAMME WORLD CONSERVATION MONITORING CENTRE (UNEP-WCMC). (2020d). *Protected Area Profile for Latin America & Caribbean from the World Database of Protected Areas, November 2020*. Retrieved December 27, 2020, from https://www.protectedplanet.net

UNITED NATIONS ENVIRONMENT PROGRAMME WORLD CONSERVATION MONITORING CENTRE (UNEP-WCMC). (2020e). *Protected Area Profile for Polar from the World Database of Protected Areas, November 2020*. Retrieved December 27, 2020, from https://www.protectedplanet.net

UNITED NATIONS ENVIRONMENT PROGRAMME WORLD CONSERVATION MONITORING CENTRE (UNEP-WCMC). (2020f). *Protected Area Profile for North America from the World Database of Protected Areas, November 2020*. Retrieved December 27, 2020, from https://www.protectedplanet.net

UNITED NATIONS ENVIRONMENT PROGRAMME WORLD CONSERVATION MONITORING CENTRE (UNEP-WCMC). (2020g). *Protected Area Profile for West Asia from the World Database of Protected Areas, November 2020*. Retrieved December 27, 2020, from https://www.protectedplanet.net

United Nations Environment Programme World Conservation Monitoring Centre (UNEP-WCMC). (2020h). *Protected Area Profile for Venezuela (Bolivarian Republic of) from the World Database of Protected Areas, November 2020.* Retrieved December 27, 2020, from https://www.protectedplanet.net

United Nations Environment Programme World Conservation Monitoring Centre (UNEP-WCMC). (2020i). *Protected Area Profile for Brazil from the World Database of Protected Areas, November 2020.* Retrieved December 27, 2020, from https://www.protectedplanet.net

United Nations Environment Programme World Conservation Monitoring Centre (UNEP-WCMC). (2020j). *Protected Area Profile for Guyana from the World Database of Protected Areas, November 2020.* Retrieved December 27, 2020, from https://www.protectedplanet.net

United Nations Environment Programme World Conservation Monitoring Centre (UNEP-WCMC). (2020k). *Protected Area Profile for Argentina from the World Database of Protected Areas, November 2020.* Retrieved December 27, 2020, from https://www.protectedplanet.net

United Nations Environment Programme World Conservation Monitoring Centre (UNEP-WCMC). (2020l). *Protected Area Profile for Uruguay from the World Database of Protected Areas, November 2020.* Retrieved December 27, 2020, from https://www.protectedplanet.net

United Nations Environment Programme World Conservation Monitoring Centre (UNEP-WCMC), & International Union for Conservation of Nature (IUCN). (2016). *Protected Planet Report 2016: How protected areas contribute to achieving global targets for biodiversity.* UNEP-WCMC and IUCN. https://portals.iucn.org/library/sites/library/files/documents/2016-051.pdf

United Nations Environment Programme World Conservation Monitoring Centre (UNEP-WCMC), & International Union for Conservation of Nature (IUCN). (2020). *Protected areas map of the world, October 2020.* Retrieved December 27, 2020, from https://www.protectedplanet.net

United Nations Environment Programme World Conservation Monitoring Centre (UNEP-WCMC), & International Union for Conservation of Nature (IUCN). (2021). *Protected Planet: The World Database on Protected Areas (WDPA), On-line, May 2021.* Retrieved September 17, 2021, from https://www.protectedplanet.net

United Nations Environment Programme World Conservation Monitoring Centre (UNEP-WCMC), International Union for Conservation of Nature (IUCN), & National Geographic Society (NGS). (2018): *Protected Planet Report 2018: Tracking progress towards global targets for protected areas.* UNEP-WCMC,

IUCN and NGS. https://livereport.protectedplanet.net/pdf/Protected_Planet_Report_2018.pdf

United Nations Environment Programme World Conservation Monitoring Centre (UNEP-WCMC), International Union for Conservation of Nature (IUCN), & National Geographic Society (NGS). (2020): *Protected Planet Live Report 2020.* UNEP-WCMC, IUCN and NGS. https://livereport.protectedplanet.net/

United Nations (UN). (1987). *Our Common Future: Report of the World Commission on Environment and Development.* UN. https://digitallibrary.un.org/record/139811

United Nations (UN). (2000). *Millennium Development Goals.* UN.

United Nations (UN). (2015a): *The Millennium Development Goals Report: 2015.* UN.

United Nations (UN). (2015b): *Transforming our world: The 2030 Agenda for Sustainable Development. A/RES/70/1.* UN.

United Nations (UN). (2020a): *The Future We Want: The United Nations We Need. Update on the Work of the Office on the Commemoration of the UN's 75th Anniversary.* UN.

United Nations (UN). (2020b). *World Statistics Pocketbook 2020 edition.* UN.

United Nations Educational, Scientific and Cultural Organization (UNESCO) (1968). *Recommendations of the Intergovernmental Conference of Experts on the Scientific Basis for Rational Use and Conservation of the Resources of the Biosphere.* UNESCO.

United Nations Educational, Scientific and Cultural Organization (UNESCO). (1971). *Final report of the first session of the International Co-ordinating Council of the Programme on Man and the Biosphere (MAB), 9-19 November 1971.* UNESCO.

United Nations Educational, Scientific and Cultural Organization (UNESCO). (1972). *Convention concerning the Protection of the World Cultural and Natural Heritage.* UNESCO.

United Nations Educational, Scientific and Cultural Organization (UNESCO). (1974). *Report of the Task Force on Criteria and Guidelines for the Choice and Establishment of Biosphere Reserves.* UNESCO.

United Nations Educational, Scientific and Cultural Organization (UNESCO). (1996). *Biosphere reserves: The Seville Strategy and the Statutory Framework of the World Network.* UNESCO. https://unesdoc.unesco.org/ark:/48223/pf0000103849.locale=en

UNITED NATIONS EDUCATIONAL, SCIENTIFIC AND CULTURAL ORGANIZATION (UNESCO). (1997). *Guidelines for establishing MAB National Committees.* UNESCO. https://unesdoc.unesco.org/ark:/48223/pf0000111527

UNITED NATIONS EDUCATIONAL, SCIENTIFIC AND CULTURAL ORGANIZATION (UNESCO). (2008). *Madrid Action Plan for Biosphere Reserves (2008-2013).* UNESCO. https://unesdoc.unesco.org/ark:/48223/pf0000163301.locale=en

UNITED NATIONS EDUCATIONAL, SCIENTIFIC AND CULTURAL ORGANIZATION (UNESCO). (2011a). *Final report of the 23rd session of the International Co-ordinating Council of the Man and the Biosphere (MAB) Programme in Dresden Radebeul, Germany, 28 June - 1 July 2011.* UNESCO. http://www.unesco.org/new/fileadmin/MULTIMEDIA/HQ/SC/pdf/SC-11-CONF-202-11_E_Final-Report.pdf

UNITED NATIONS EDUCATIONAL, SCIENTIFIC AND CULTURAL ORGANIZATION (UNESCO). (2011b). *Statutes of the International Advisory Committee for Biosphere Reserves.* UNESCO.

UNITED NATIONS EDUCATIONAL, SCIENTIFIC AND CULTURAL ORGANIZATION (UNESCO). (2012a). *Final report of the 24th session of the International Co-ordinating Council of the Man and the Biosphere (MAB) Programme in UNESCO Headquarters, Paris, 9-13 July 2012.* UNESCO. https://unesdoc.unesco.org/ark:/48223/pf0000378686.locale=en

UNITED NATIONS EDUCATIONAL, SCIENTIFIC AND CULTURAL ORGANIZATION (UNESCO). (2012b). *Records of the General Conference, 36th session, Paris, 25 October - 10 November 2011: Volume 1, Resolutions.* UNESCO.

UNITED NATIONS EDUCATIONAL, SCIENTIFIC AND CULTURAL ORGANIZATION (UNESCO). (2013). *Final report of the 25th session of the International Co-ordinating Council of the Man and the Biosphere (MAB) Programme in UNESCO Headquarters, Paris, 27-30 May 2013.* UNESCO. https://unesdoc.unesco.org/ark:/48223/pf0000378687.locale=en

UNITED NATIONS EDUCATIONAL, SCIENTIFIC AND CULTURAL ORGANIZATION (UNESCO). (2014). *Final report of the 26th session of the International Co-ordinating Council of the Man and the Biosphere (MAB) Programme in Jönköping, East Vättern Scarp Landscape Biosphere Reserve, Sweden, 10 - 13 June 2014.* UNESCO. https://unesdoc.unesco.org/ark:/48223/pf0000244796.locale=en

UNITED NATIONS EDUCATIONAL, SCIENTIFIC AND CULTURAL ORGANIZATION (UNESCO). (2016a). *Lima Action Plan for UNESCO's Man and the Biosphere (MAB) Programme and its World Network of Biosphere Reserves (2016-2025).* UNESCO. http://www.unesco.org/new/fileadmin/MULTIMEDIA/HQ/SC/pdf/Lima_Action_Plan_en_final_01.pdf

United Nations Educational, Scientific and Cultural Organization (UNESCO). (2016b). *Final report of the 28th session of the International Co-ordinating Council of the Man and the Biosphere (MAB) Programme in Lima, Peru, 18 - 19 March 2016.* UNESCO. https://unesdoc.unesco.org/ark:/48223/pf0000378679.locale=en

United Nations Educational, Scientific and Cultural Organization (UNESCO). (2017a). *Final report of the 29th session of the International Co-ordinating Council of the Man and the Biosphere (MAB) Programme in UNESCO Headquarters, Paris, 12-15 June 2017.* UNESCO. https://unesdoc.unesco.org/ark:/48223/pf0000253591

United Nations Educational, Scientific and Cultural Organization (UNESCO). (2017b). *Periodic Review Process.* Retrieved December 12, 2020, from http://www.unesco.org/new/en/natural-sciences/environment/ecological-sciences/biosphere-reserves/periodic-review-process/

United Nations Educational, Scientific and Cultural Organization (UNESCO). (2017c). *Revalidation Process of UNESCO Global Geoparks.* Retrieved January 1, 2021, from http://www.unesco.org/new/en/natural-sciences/environment/earth-sciences/unesco-global-geoparks/revalidation-process/

United Nations Educational, Scientific and Cultural Organization (UNESCO). (2017d). *Lake Bosomtwe.* Retrieved January 24, 2021, from http://www.unesco.org/new/en/natural-sciences/environment/ecological-sciences/biosphere-reserves/africa/ghana/lake-bosomtwe/

United Nations Educational, Scientific and Cultural Organization (UNESCO). (2018). *Final report of the 30th session of the International Co-ordinating Council of the Man and the Biosphere (MAB) Programme in Palembang, South Sumatra Province, Indonesia, 23-28 July 2018.* UNESCO. http://www.unesco.org/new/fileadmin/MULTIMEDIA/HQ/SC/pdf/FINAL_30_MAB_ICC_REPORT_en.pdf

United Nations Educational, Scientific and Cultural Organization (UNESCO). (2019). *Final report of the 31st session of the International Co-ordinating Council of the Man and the Biosphere (MAB) Programme in Paris, France, 17-21 June 2019.* UNESCO. https://unesdoc.unesco.org/ark:/48223/pf0000370193

United Nations Educational, Scientific and Cultural Organization (UNESCO). (2020a). *Item 6 of the provisional agenda: Report of the Secretary of the MAB Programme. 32nd session of the ICC MAB Programme, online session 27-28 October 2020.* UNESCO.

United Nations Educational, Scientific and Cultural Organization (UNESCO). (2020b). *Item 9 of the provisional agenda: Proposals for New Biosphere Reserves and Extensions/ Modifications/ Renaming to Biosphere Reserves that are part of the World Network of Biosphere Reserves (WNBR).* 32nd session of the ICC MAB Programme, online session 27-28 October 2020. UNESCO.

United Nations Educational, Scientific and Cultural Organization (UNESCO). (2020c). *World Network of Biosphere Reserves*. Retrieved December 19, 2020, from https://en.unesco.org/biosphere/wnbr

United Nations Educational, Scientific and Cultural Organization (UNESCO). (2020d). *Man and the Biosphere (MAB) Programme: MAB Governance*. Retrieved January 01, 2021, from https://en.unesco.org/mab/governance

United Nations Educational, Scientific and Cultural Organization (UNESCO). (2020e). *List of UNESCO Global Geoparks (UGGp)*. Retrieved January 3, 2021, from http://www.unesco.org/new/en/natural-sciences/environment/earth-sciences/unesco-global-geoparks/list-of-unesco-global-geoparks/

United Nations Educational, Scientific and Cultural Organization (UNESCO). (2022). *Final report of the 33rd session of the International Co-ordinating Council of the Man and the Biosphere (MAB) Programme in Abuja, Nigeria (Hybrid meeting), 13-17 September 2022*. UNESCO. https://en.unesco.org/sites/default/files/sc-21_conf233_15-mab_icc_report_eng_0.pdf

United Nations Educational, Scientific and Cultural Organization (UNESCO) World Heritage Centre (WHC). (2019). *Operational Guidelines for the Implementation of the World Heritage Convention*. UNESCO. https://whc.unesco.org/document/178167

United Nations Educational, Scientific and Cultural Organization (UNESCO) World Heritage Centre (WHC). (2020a). *Periodic Reporting*. Retrieved January 3, 2021, from https://whc.unesco.org/en/periodicreporting/

United Nations Educational, Scientific and Cultural Organization (UNESCO) World Heritage Centre (WHC). (2020b). *World Heritage List*. Retrieved January 3, 2021, from http://whc.unesco.org/en/list/

United Nations Development Programme (UNDP). (2020). *Human Development Report 2020: The next frontier. Human development and the Anthropocene*. UNDP. https://hdr.undp.org/sites/default/files/hdr_2020_overview_english.pdf

U.S. Man and the Biosphere Program. (1994). *Strategic Plan for the U.S. Biosphere Reserve Program*. Department of State Publication. http://npshistory.com/publications/mab/OPN_BR_12.pdf

U.S. Man and the Biosphere Program. (1995). *Biosphere Reserves in Action: Case Studies of the American Experience*. Department of State Publication. https://doi.org/10.2172/771250

Van Hoven, W., & Bashir Nimir, M. B. (2004). Recovering from conflict: the case of Dinder and other national parks in Sudan. *Parks, 14*(1), 26-34. https://www.iucn.org/sites/dev/files/import/downloads/14_1.pdf

VERWEIJEN, J., & MARIJNEN, E. (2018). The counterinsurgency/conservation nexus: guerilla livelihoods and the dynamics of conflict and violence in Virunga National Park, Democratic Republic of the Congo. *The Journal of Peasant Studies, 48*(2), 300-320. https://doi.org/10.1080/03066150.2016.1203307

WALDRON, A., ADAMS, V., ALLAN, J., ARNELL, A., ASNER, G., ATKINSON, S., BACCINI, A., BAILLIE, J., BALMFORD, A., BEAU, J. A., BRANDER, L., BRONDIZIO, E., BRUNER, A., BURGESS, N., BURKART, K., BUTCHART, S., BUTTON, R., CARRASCO, R., CHEUNG, W., CHRISTENSEN, V., CLEMENTS, A., COLL, M., DI MARCO, M., DEGUIGNET, M., DINERSTEIN, E., ELLIS, E., EPPINK, F., ERVIN, J., ESCOBEDO, A., FA, J. FERNANDES-LLAMAZARES, A., FERNANDO, S., FUJIMORI, S., FULTON, B., GARNETT, S., GERBER, J., GILL, D., GOPALAKRISHNA, T., HAHN, N., HALPERN, B., HASEGAWA, T., HAVLIK, P., HEIKINHEIMO, V., HENEGHAN, R., HENRY, E., HUMPENODER, F., JONAS, H., JONES, K., JOPPA, L., JOSHI, A. R., JUNG, M., KINGSTON, N., KLEIN, C., KRISZTIN, T., LAM, V., LECLERE, D., LINDSEY, P., LOCKE, H., LOVEJOY, T. E., MADGWICK, P., MALHI, Y., MALMER, P., MARON, M., MAYORGA, J., VAN MEIJL, H., MILLER, D., MOLNAR, Z., MUELLER, N., MUKHERJEE, N., NAIDOO, R., NAKAMURA, K., NEPAL, P., NOSS, R. F., O'LEARY, B., OLSON, D., PALCIOS ABRANTES, J., PAXTON, M., POPP, A., POSSINGHAM, H., PRESTEMON, J., RESIDE, A., ROBINSON, C., ROBINSON, J., SALA, E., SCHERRER, K., SPALDING, M., SPENCELEY, A., STEENBECK, J., STEHFEST, E., STRASSBORG, B., SUMAILA, R., SWINNERTON, K., SZE, J., TITTENSOR, D., TOIVONEN, T., TOLEDO, A., NEGRET TORRES, P., VAN ZEIST, W., VAUSE, J., VENTER, O., VILELA, T., VISCONTI, P., VYNNE, C., WATSON, R., WATSON, J., WIKRAMANAYAKE, E., WILLIAMS, B., WINTLE, B., WOODLEY, S., WU, W., ZANDER, K., ZHANG, Y., & ZHGNA, Y. P. (2020). *Protecting 30% of the planet for nature: costs, benefits and economic implications.* Working paper analysing the economic implications of the proposed 30% target for areal protection in the draft post-2020 Global Biodiversity Framework. https://www.conservation.cam.ac.uk/files/waldron_report_30_by_30_publish.pdf

WATSON, J. E. M., DUDLEY, N., SEGAN, D. B., & HOCKINGS, M. (2014). The performance and potential of protected areas. *Nature, 515*, 67-73. https://doi.org/10.1038/nature13947

WELLS, P. (1996). The social role of protected areas in the new South Africa. *Envionmental Conservation, 23*(4), 322-331. https://www.jstor.org/stable/44519315

WINKLER, K. J. (2019). The implementation of the conceptual shift of conservation: Pathways of three German UNESCO biosphere reserves. *Ecosystems and People, 15*(1), 173-180. https://doi.org/10.1080/26395916.2019.1617351

WISSENSCHAFTLICHER BEIRAT DER BUNDESREGIERUNG GLOBALE UMWELTVERÄNDERUNGEN (WBGU). (2016). *Der Umzug der Menschheit: Die transformative Kraft der Städte.* WBGU. https://issuu.com/wbgu/docs/wbgu_hg2016-hoch?e=37591641/68732842

WOODLEY, S., RAO, M., MACKINNON, K., SANDWITH, T., & DUDLEY, N. (2021). Speaking a common language on what should count for protecting 30 per cent by 2030? *Parks, 27.2*, 9-14. https://doi.org/10.2305/IUCN.CH.2021.PARKS-27-2SE.en

WORLD ECONOMIC FORUM. (2020): *The Global Risks Report 2020: 15th Edition*. World Economic Forum. https://www3.weforum.org/docs/WEF_Global_Risk_Report_2020.pdf

WYSS CAMPAIGN FOR NATURE. (2019). *Fact Sheet: A Plan to Protect at least 30 Percent of our Planet by 2030*. Retrieved September 17, 2021, from https://www.wysscampaign.org/news/2019/1/22/fact-sheet-a-plan-to-protect-at-least-30-percent-of-our-planet-by-2030

XIANPU, W. (1994). East Asia. In: J. McNeely, J. Harrison, & P. Dingwall (Eds.), *Protecting Nature: Regional Reviews of Protected Areas*. IUCN, 157-175.

XUN, Z., & WANG, M. (2002). National geoparks initiated in China: Putting geoscience in the service of society. *Episodes, 25*(1), 33-37. http://doi.org/10.18814/epiiugs/2002/v25i1/005

YELLOWSTONE NATIONAL PARK PROTECTION ACT, 42nd Cong., 2d Sess. (1872) (enacted). https://www.archives.gov/legislative/features/national-park-service/yellowstone.html

YOSEMITE ACT, 38th Cong., 1st Sess. (1864) (enacted). https://www.archives.gov/legislative/features/national-park-service/yosemite.html

ZIMMERER, K. S. (2006). Cultural ecology: at the interface with political ecology - the new geographies of environmental conservation and globalization. *Progress in Human Geography, 30*(1), 63-78. https://doi.org/10.1191%2F0309132506ph591pr

ZIMMERER, K. S., GALT, R. E., & BUCK, M. V. (2004). Globalization and Multi-spatial Trends in the Coverage of Protected-Area Conservation (1980-2000). *AMBIO: A Journal of the Human Environment, 33*(8), 520-529.

Annex

Annex 1: Legend of detailed land cover classification used in maps in this dissertation

Colour	Value	Land cover classification
	10	Cropland, rainfed
	11	Herbaceous cover
	12	Tree or shrub cover
	20	Cropland, irrigated or post-flooding
	30	Mosaic cropland (>50%) / natural vegetation (tree, shrub, herbaceous cover) (<50%)
	40	Mosaic natural vegetation (tree, shrub, herbaceous cover) (>50%) / cropland (<50%)
	50	Tree cover, broadleaved, evergreen, closed to open (>15%)
	60	Tree cover, broadleaved, deciduous, closed to open (>15%)
	61	Tree cover, broadleaved, deciduous, closed (>40%)
	62	Tree cover, broadleaved, deciduous, open (15-40%)
	70	Tree cover, needleleaved, evergreen, closed to open (>15%)
	71	Tree cover, needleleaved, evergreen, closed (>40%)
	72	Tree cover, needleleaved, evergreen, open (15-40%)
	80	Tree cover, needleleaved, deciduous, closed to open (>15%)
	81	Tree cover, needleleaved, deciduous, closed (>40%)
	82	Tree cover, needleleaved, deciduous, open (15-40%)
	90	Tree cover, mixed leaf type (broadleaved and needleleaved)
	100	Mosaic tree and shrub (>50%) / herbaceous cover (<50%)
	110	Mosaic herbaceous cover (>50%) / tree and shrub (<50%)
	120	Shrubland
	121	Shrubland evergreen
	122	Shrubland deciduous
	130	Grassland
	140	Lichens and mosses

	150	Sparse vegetation (tree, shrub, herbaceous cover) (<15%)
	151	Sparse vegetation (tree, shrub, herbaceous cover) (<15%)
	152	Sparse shrub (<15%)
	153	Sparse herbaceous cover (<15%)
	160	Tree cover, flooded, fresh or brackish water
	170	Tree cover, flooded, saline water
	180	Shrub or herbaceous cover, flooded, fresh/saline/brackish water
	190	Urban areas
	200	Bare areas
	201	Consolidated bare areas
	202	Unconsolidated bare areas
	210	Water bodies
	220	Permanent snow and ice

Annex 2: Questionnaire for expert interviews

Julius-Maximilians-
UNIVERSITÄT
WÜRZBURG

Institut für Geographie und Geologie
LEHRSTUHL FÜR GEOGRAPHIE UND
REGIONALFORSCHUNG

Questionnaire PhD-project

(Manuel Engelbauer, M.Sc.)

A global assessment of UNESCO Biosphere Reserves as models for sustainable development: Recent challenges in the designation process, quality management and interlinkages with other UNESCO/PA labels

Opening question

- *What are the biggest challenges UNESCO Biosphere Reserves are facing at the moment? Globally and in your country?*

Designation process

- *What main challenges and problems resulted out of the changes in the designation process from Sevilla/Madrid/Lima?*
- *How are Biosphere Reserves integrated in the legislation of your country?*
- *Do you use the name "biosphere reserves" in your country?*

Quality management

- *How does the quality management of biosphere reserves work in your country?*
- *What do you think about the periodic review system?*
- *What possibilities could be used to improve the evaluation process? Are there any specific actors who could be involved?*
- *What could be the role of the regional or thematic MAB networks?*
- *Why was it necessary to implement the exit strategy?*
- *What are the main problems of the execution of the Exit-Strategy?*

Interlinkages with other PA labels

- *UNESCO-Labels: Geopark vs. World Heritage Sites vs. Biosphere Reserves: is it a supplement system of labels or are they competing against each other?*
- *Are there any challenges between the UNESCO labels in your country?*

Models for sustainable development

- *If Biosphere Reserves are seen as model regions, then how many do we need? Are 686 too much or not enough?*